SKILLINGERS
OF
BRIGHTLINGSEA

SKILLINGERS OF BRIGHTLINGSEA

Essex's Offshore Oyster-Catchers

SEAN O'DELL

The History Press

For Mum and Dad

The Beach, Brightlingsea.

First published 2009

The History Press
The Mill, Brimscombe Port
Stroud, Gloucestershire, GL5 2QG
www.thehistorypress.co.uk

© Sean O'Dell, 2009

The right of Sean O'Dell to be identified as the Author
of this work has been asserted in accordance with the
Copyrights, Designs and Patents Act 1988.

British Library Cataloguing in Publication Data.
A catalogue record for this book is available from the British Library.

ISBN 978 0 7524 5089 6

Typesetting and origination by The History Press
Printed in Great Britain

CONTENTS

ACKNOWLEDGEMENTS

Firstly I should like to express my gratitude to my wife Nicola and my children Patrick and Kate for their unwavering support during the time that I was engaged in research for this project – a busy time in all our lives. Also, thanks to Dr David Dymond for his guidance, encouragement and enthusiasm during all stages of its preparation, right through to publication. Thanks are also due to Dr Evelyn Lord for her constant support and commitment throughout; to Amy Rigg at The History Press for her continued help, advice and support; to Margaret Stone of the Brightlingsea Museum for arranging full access to the archives, for always being available at short notice to open up Brightlingsea Museum and reading room to me, and for permission to reproduce some of their remarkable collection of old photographs; to the staff of the Essex Record Office at Chelmsford (and Colchester, prior to closure) for their courteous and invaluable assistance; to Alan Williams and Peter Allen for sharing with me their intimate and extensive knowledge of Colne fishing smacks; to Steve and Phil Munro for their invaluable help with the proof-reading; and finally to Mark Donachy, William Bortrick and Dr John Walter at Cambridge for their encouragement, laughs and lots of cups of tea.

LIST OF ABBREVIATIONS

BPM: *Brightlingsea Parish Magazine.*
CUL: Cambridge University Library.
ERO: Essex Record Office.
ERO/COL: Essex Record Office, Colchester Branch.
NA: National Archives.
PP(O/L): Parliamentary Papers On-Line, accessed at CUL.

Please note:

For printed sources the place of publication is London unless otherwise stated.
The technical terms used in this study are explained in the glossary.

FOREWORD

For many centuries Brightlingsea was a scattered farming community, but its possession of 'hard' or natural landing places on the tidal creek always gave it a secondary economic interest in fishing. In the nineteenth century growth in the demand for oysters, particularly from London, led to the rapid transformation of three scattered hamlets at the southern end of the parish into a busy port and a genuine town. In the second half of the century its population more than doubled; domestic housing, industrial premises and public buildings shot up; secular and religious institutions were founded (typical of a Victorian town); a railway line was installed giving it direct access to London, and links were established with other parts of England and Europe. However, when its fundamental trade in offshore oysters collapsed at the end of the nineteenth century, Brightlingsea's economy slumped, its growth froze and it failed to achieve its ambition of becoming a self-governing borough.

This study comes out of a dissertation submitted for the degree of MSt at Cambridge in 2007. As Sean O'Dell's former supervisor I am delighted to see his work now appear in book form, and heartily commend it to a wider readership. The following pages are an important contribution to urban history, but at the lower end of the scale, for they concentrate on genuine urban development of a limited and temporary kind.

David Dymond

CHAPTER ONE

Brightlingsea: Then and Now

In January 1995 Brightlingsea, a small port on the north coast of Essex, witnessed desperate and unprecedented scenes as police struggled with protestors attempting to stop livestock merchants exporting live animals from the town's wharf. The townsfolk were almost unanimously opposed to this trade operating on their doorstep, and were ultimately successful in stopping the practice. This was not the first time that Brightlingsea had been significantly affected by external influences: over 150 years earlier a process began that would bring about permanent change to the community, shaping the town that stands today. The purpose of this book, therefore, is to consider how and why the town developed so rapidly after 1800, and in particular to look at Brightlingsea's short-lived but significant economic boom centred on the maritime trade in the nineteenth and early twentieth centuries, and to relate these findings to the greater scheme and national experience.

The town known today as Brightlingsea was mentioned in Domesday as 'Brictriceseia', a manor of ten hides held by King Harold.[1] The geographical location of the settlement has always been advantageous for fishing and in particular for the gathering of oysters. The early maritime significance of the 'town' is confirmed in a contemporary manuscript which details ships that were provided for defence against the Armada: 'All the menes names that belongethe to the barck Parnell of Briclesea sarving Sir Fraunsses in the yeare of our lorde god 1589.' This vessel of eighty tons was said to be one of the two largest at Brightlingsea at this time.[2]

By 1801 the 'town' was a relatively modest one with a population of just 807 persons.[3] Yet by 1872 the community is described in a gazetteer as a parish of 3,560 acres on the estuary of the River Colne (eight miles south-east of Colchester) with a population of 2,585 and a large trade in the fishing of sprats and oysters.[4] By the beginning of the First World War the town had undergone considerable development. It is this period of change that will be the focus of this study.[5]

Map of Brightlingsea within the UK.

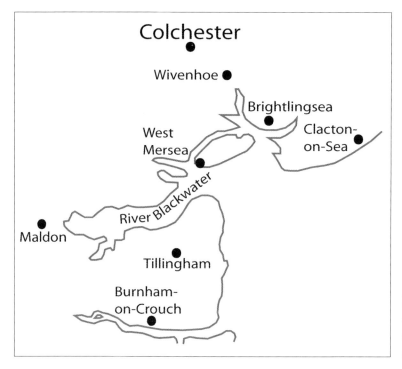

Brightlingsea and the surrounding area.

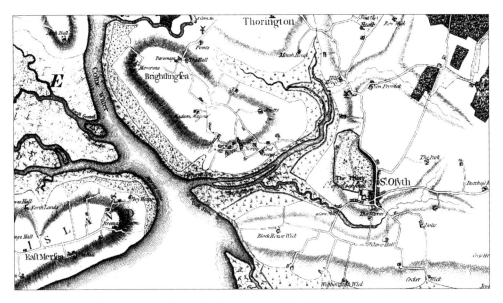

Brightlingsea and Saint Osyth from the Chapman and André map, 1777.

From early times Brightlingsea's fortunes were linked to inshore oyster grounds, and by the eighteenth century these had been developed into a busy local industry. In 1797 Major Thomas Reynolds was charged with the task of surveying the area, and he provided the following description:

> About half a mile up Brightlingsea Creek, vessels of 3 to 4ft. draft may ground on a fine hard beach at almost all times of the tide. Brightlingsea is a large straggling village, inhabited by the oyster dredgers and their families. Fifty boats are employed in the fishery, each worked on average by two men and a boy. These people have formed some regulations among themselves, and a Mr Tabor, a principle proprietor in the oyster trade, acts as mayor.[6]

As the nineteenth century saw an increase in demand for oysters, so the inshore oyster beds known as 'layings' struggled to satisfy the growing market in the 1790s. Some local fishermen dredged for offshore oysters much farther out to sea, but not on a regular basis.[7] The coarse, more distinctively flavoured oysters found in offshore grounds were of a different species from their cultivated counterparts farmed in sheltered inshore layings, but were nevertheless equally popular with consumers from all social strata during the nineteenth century.[8]

Harvesting offshore varieties was a far more arduous business, however, requiring considerably more substantial vessels than the smaller estuary smacks used in the sheltered waters close to home. Crews had to make longer voyages to more distant destinations, often in foul weather, in order to obtain their catch. Consequently the activity was far more hazardous than most others within the fishing industry at that time. As previously mentioned, the smack crews ventured offshore on an occasional basis for many years, but it was not until the demand for oysters began to increase during the early to mid-nineteenth century that the activity can be said to be a separate phenomenon, demanding the adaptation of existing vessels and the construction of new ones for skippers who were working solely on the offshore grounds. These new offshore smacks were generally much larger and of heavier construction, in order to withstand heavy seas. The increased size enabled a larger catch to be carried – an important factor for the offshore traders. The provision of a special tank, or 'wet well', onboard allowed the catch to be kept alive and fresh during the longer voyages. These technical complexities are discussed in more depth in chapter two. The offshore crews were also able to assist in the re-stocking of the inshore layings using brood oysters from other inshore grounds further away. The heavier offshore smacks were, like their smaller inshore cousins, still remarkably fast under sail. This was a useful advantage as local crews of the areas into which they ventured did not always welcome the Brightlingsea fishermen.[9]

With these larger vessels and a healthy demand for their catch, Brightlingsea crews roamed far and wide in search of oysters: to the west coast and Cornwall, the Channel Islands, French coastal waters and even as far north as Scotland. The most notorious destination, however, was the Terschelling Banks off the Dutch coast where, in severe weather, several smacks and crew members were lost. The offshore smacks owed their nickname 'Skillingers' to a corruption of 'Terschelling'. By now, around the last quarter of the nineteenth century, the offshore oyster industry had come of age.[10]

Brightlingsea, with its safe harbour and easy access to the open sea, was well situated at the centre of this new branch of the oyster industry. An important factor has to be the location and development of the Aldous Shipyard on the waterfront, specialising in the designing, converting and building of larger offshore smacks, in addition to yachts. Another advantage was the fact that the area had long been at the forefront of the supply of oysters, and an infrastructure for the disposal of the catch was already in place. The town's association with oysters was known throughout recorded history and has been frequently mentioned in documents dealing with rights to oyster layings in the Colne Estuary, part of which fell under the control of nearby Colchester.[11] Brightlingsea also has the title of Cinque Port Liberty, being a limb of Sandwich, and the only member of that confederation in Essex.[12]

Geographically speaking Brightlingsea is virtually an island, and in earlier times was referred to as such. Today the B1029 is still the only road entering the town from the outside world. The parish boundary generally follows the natural lines of tidal creeks and streams, and this has meant that the town has had to develope in a fairly isolated environment.[13] Brightlingsea has, therefore, in many ways, looked to the sea as its 'front door' and its hinterland as the 'tradesman's entrance'.

Work by Ravenstein and others has raised several fundamental questions worthy of investigation. He suggested that increasing trade and industrial activity within a town or region can provide a 'pull' factor for the inward migration of workers and families who at the same time may experience a 'push' from areas where trade or industry is in decline.[14] During the nineteenth century Brightlingsea's growing offshore oyster and maritime trade could therefore have attracted existing smack skippers and crews, or, as Long suggests, 'the cream of the rural labour market crop': potential apprentices looking for new work?[15] As word of successful and lucrative hauls of the ever more popular oyster from offshore grounds spread through fishing communities, how many young men saw an opportunity for improving their livelihoods by moving into this area? Brightlingsea, as has been suggested, was traditionally seen as a centre for the oyster trade and would have exerted the 'pull' factor described above. Young men who would otherwise have relied on agricultural labouring in the surrounding region may have seen more lucrative and adventurous opportunities onboard the oyster boats. But was the offshore oyster industry the only activity fuelling the boom? Opportunities for more specialised trades may also have arisen, for example among chandlers and rope makers and, of course, manning the racing yachts during the summer months.

Many coastal towns in England owe their former prosperity to the fishing industry, but how typical was Brightlingsea's experience during this time? Was the town's growth paralleled in other coastal communities? Did other local villages or towns not engaged in the fishing industry experience similar growth and development?

Statistics from nineteenth-century census returns do indeed show a strong increase in the town's population. In 1801 a figure of 807 persons is recorded, which by 1851 had increased to 1,852 persons shown as resident. By 1891 this figure had increased further to 3,920, and by 1901 the population had, quite remarkably, reached 4,501, where it plateaued.

It can be seen from the chart on page 14 that a distinct increase in population occurred between 1851 and 1901, coinciding with the growth of the offshore oyster fishing industry and the town's general maritime activities.

Brightlingsea has had a healthy number of books and pamphlets published dealing with most aspects of its past. These range from a comprehensive history of the town produced by local medical practitioner Dr Dickin,[16] and a more recent study of the town's Cinque Port heritage, to smaller commemorative and sometimes nostalgic booklets dealing with aspects of the town's general maritime heritage. This heritage is also covered in some detail by two authors who have studied the subject more widely along the east coast. The works of Hervey Benham and John Leather cover many aspects of seafaring along the East Anglian coast, particularly from the mid-nineteenth century onward.[17] They include detailed information about the fishing industries (including the Brightlingsea trade) and yacht racing, in terms of the methods employed, the vessels used and the daily working lives of the fishermen and mariners.

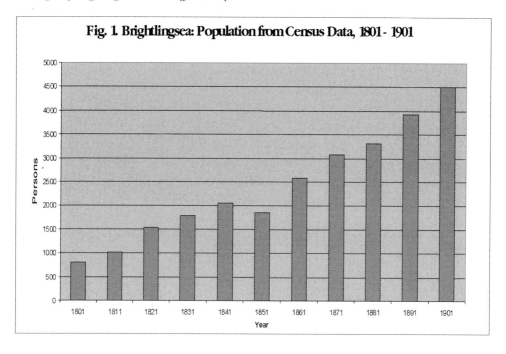

Fig. 1 Brightlingsea: Population from Census Data, 1801 - 1901

As far as Brightlingsea is concerned, however, one important area seems to have remained unexplored. As has been suggested, the offshore oyster-fishing industry that developed as a result of increasing demand (which the inshore industry could apparently not fulfil) was relatively short-lived but highly significant. Whilst this activity has been described by the two authors mentioned earlier, the effect that it and its allied trades and activities had upon the town demographically, economically and socially has remained largely unexplored. There are of course works that deal with these issues in other localities, and more generally with the regional and national fishing industries. Nevertheless, there appears to be a gap in the local historiography, and given the geographical location and historical significance of Brightlingsea as a maritime centre, a 'micro-historical' study at a lower level which in turn relates any findings to the regional as well as the national picture is much needed. The secondary works mentioned above provide a background to the subject and a starting point from which to exploit a range of primary sources which include census enumerations, various parochial and manorial records, a remarkable parish magazine, maps, early photographs and parliamentary reports and accounts.

Today Brightlingsea is a thriving community, home to a busy yachting and pleasure-boating scene during the summer months, with a number of local businesses at the centre of this activity. Much redevelopment of the waterfront for flats and housing has taken place in recent years, and the area of the former James and Stone Shipyard has been extensively re-developed. The town still has a lively high street with a range of

Brightlingsea Museum and reading rooms.

shops and services, and the church and chapels are also still very much at the heart of the community, as they were during the busy period of expansion in the nineteenth century.

The waterfront may look very different today compared to the busy commercial fishing, oyster dredging and yachting days of the late nineteenth century, but take a walk along the waterfront past the Hard-Masters hut toward the Aldous Heritage Dock where the Colne Smack Preservation Society members work on their beloved vessels and strike up a conversation with someone working ashore or afloat, or visit the excellent museum in town, and you will quickly realise that Brightlingsea's remarkable maritime heritage is still very much at the heart of this charming Essex town.

The new marina and flats, 2009.

Smacks at the dock.

The Aldous Heritage Smack Dock.

CHAPTER TWO

The Oyster Industry and Maritime Trades

'True, true oh oyster! Thou art the best beloved of the loved! After eating oysters we feel joyous, light and agreeable – yes, one might say, fabulously well.'[18]

The sentiment expressed above epitomises attitudes towards the consumption of oysters amongst the middle and upper classes in England as the nineteenth century progressed. And yet, at the beginning of the century oysters were abundant and popular with all strata of society (as they had been in the past); indeed, they were a staple food of the poorer classes in the areas around England's extensive coastline, and particularly in London.[19] Charles Dickens, writing in 1837, reflected this:

> 'It's a wery remarkable circumstance, sir' said Sam, 'that poverty and oysters seem to go together.'
>
> 'I don't understand you, Sam,' said Mr. Pickwick.
>
> 'What I mean, sir,' said Sam, 'is, that the poorer a place is, the greater call there seems to be for oysters. Look here, sir; there's an oyster stall to every half-dozen houses. The street's lined with 'em. Blessed if I don't think that ven a man's wery poor, he rushes out of his lodgings, and eats oysters in reg'lar desperation.'[20]

With the development of rail connections from the cities and towns to the coast, inland markets began to see the arrival of more and more fresh seafood. This was aided by the decreasing cost of ice in which the produce was packed to keep it fresh for longer.[21] The generally held belief that oysters are particularly beneficial to health and well-being also appeared to gain currency around this time, which undoubtedly promoted their popularity with the middle and upper classes. Increasing availability of fresh produce and its growing popularity as a 'fashionable' foodstuff set the scene for

Advertisement extolling the
virtues of oysters for health.

a mid-nineteenth-century boom in oyster consumption: Hall observed in 1878 that
'of the quantity of oysters consumed in London we cannot give even an approximate
guess. It must amount to millions of bushels.'[22]

This considerable increase in demand for oysters was apparent at Billingsgate, the
area of London known for selling fish, which developed in the eighteenth century
into a large and busy market place.[23] Writing in *London Labour and the London Poor* in
1851, Henry Mayhew gave a vivid description of the scene near Billingsgate, noting
the large numbers of vessels laden with oysters that were tied-up alongside the wharf.
He also observed that the costermongers nicknamed this area 'oyster street', and that
it teemed with traders and the general public keen to buy the fresh native oysters.[24]
In addition, Hall noted: 'Billingsgate in the oyster season is a sight and a caution.
Boats coming in loaded; porters struggling with baskets and sacks; early loungers
looking on'.[25] By the mid-1800s the trade at Billingsgate was enormous: 495,896,000
oysters are recorded as being sold there in 1864. Mayhew reports that around
124,000,000 were sold annually by the costermongers at four-a-penny, which means
that £12,916 13s 4d was spent on oysters each year on the London streets alone.[26]
Mayhew describes this activity for us in rich detail in *London Labour and the London
Poor*, in a passage entitled 'Of Oyster Selling in the Streets', which is transcribed in
Appendix One.

Brightlingsea's proximity to the London market (about seventy miles by rail, or
ninety nautical miles) meant that the town's mariners would undoubtedly benefit
from this increasing demand. Although Colchester Borough controlled a large part
of the Colne Estuary and its associated layings, Brightlingsea's mariners had their own
common grounds at the seaward end of the estuary and around the harbour. As demand
increased they would expand their dredging activities further afield, ultimately
venturing offshore on a far more regular basis. Existing accounts from the local oyster
merchants are scarce, but those that have survived from the latter part of the nineteenth

Smacks ashore at Point Clear.

century do in fact show all the signs of a brisk trade.[27] Genuine details of crew's wages are also scarce; hauls of oysters were generally sold ashore for the prevailing price, and the crew's reward would therefore depend on market conditions.[28]

Oyster dredging was, in earlier times, essentially an inshore activity. Before increasing demand began to put pressure upon supplies (and the cultivation of oysters in beds and layings became necessary) they could be dredged from their natural habitats on the gravelly estuary beds. Early dredging equipment consisted of perforated leather 'bags' with weighted wooden beams at the open end that allowed the dredges to be dragged along the bed collecting the oysters as they went. This simple dredge evolved into a more sophisticated design, with rope netting and eventually a metal ring mesh replacing the leather bag. Different sizes were used for various applications.[29] Offshore dredges were, of course, much larger and capable of operating at greater depths and gathering much bigger hauls. They were, therefore, much heavier and harder to handle than their smaller inshore counterparts, and they would be hauled aboard time after time by the crew during their long shift, day or night, in all weathers.

During the nineteenth century relatively small sailing vessels known as 'bumkins' were used to dredge in the sheltered inshore waters of the Colne Estuary as it gradually wound its way towards Wivenhoe and Colchester.[30] Many of Brightlingsea's oyster crews, being independent of the Colchester-controlled Colne Oyster Fishery, worked the 'common grounds' locally and further afield and therefore needed a more seaworthy vessel. These vessels, or 'smacks', were fast under sail, could be managed by a

Smacks afloat, 1906.

244CK in the creek.

Above and below: Launching the restored CK21.

crew of three and carry a reasonable haul of oysters. Speed was an essential component as oysters needed to be kept fresh before being sold on.

The 'second-class' smacks described above were up to 45ft in length and 13ft 6in in the beam. They were built with frames of oak, with planking and decking of pitch pine. Elm was used for the garboard strake, a particularly vulnerable plank just above the keel, as it was good in water and held a nail well. The expansive decks carried no superstructure as work space was precious; accommodation (such as it was) was in the forepeak where a small stove and cots nestled around the sail locker. The rest of the space below the deck was used for storing the catch. The mast was of solid pitch pine, generally the same length as the deck, and carried Admiralty-grade flax sails in a gaff-rig.[31]

Increasing demand in the mid- to late nineteenth century forced crews to work further afield, and trips were extended to around twelve days for Terschelling and up to two or more weeks for further distances.[32] The large offshore smack was developed with a 'wet well' for keeping the haul alive during these extended voyages. The wet well was in effect a tank built into the vessel that was partially open to the sea, allowing fresh sea-water to flow constantly over the oysters kept within.

The large 'first-class' smacks were of the same basic construction as the 'second-class' smacks, but were much heavier (to withstand the heavy winter seas and long voyages), had more accommodation below decks and were rigged as cutters with up to 70ft in length and 15ft in the beam.[33]

The smacks *Pioneer, Lady Olive* and *Excellent* from Point Clear, *c.*1904.

White's directory for 1848 informs us that annually between February and March Brightlingsea sent sixty smacks to Jersey and the south-west channel to dredge for oysters. The directory also claims that about 300 inhabitants were licensed to dredge for oysters in the Colne Estuary and nearby creeks by the Corporation of Colchester, and that around 160 smacks were involved.[34] According to a report to the Board of Trade in 1867, Brightlingsea's sea-going oyster fleet was by now the largest on the north Essex coast, comprising of some seventy-four vessels (totalling 819 tons) crewed by 212 men and boys.[35] The continually increasing demand for oysters caused Brightlingsea's growing population of mariners to demand locally built vessels capable of making extended voyages for longer periods, and the waterfront yard of the Aldous family in particular was able to accommodate this need, building thirty-six of the new larger smacks between 1857 and 1867.[36] As the century wore on so the fleet grew, and the construction of the largest first-class smacks commenced; the Aldous yard built the 34-ton *Christabel* (CK66) in 1869 and the 39-ton, 63ft *Excellent* (CK30) in 1883, to name but two.[37] James Aldous had bought the site by the waterfront back in 1833, and thereafter set about moving away from his business as a builder and contractor toward work at the shipyard. By the middle of the century the yard's reputation for building fine vessels was well established, and the knowledge and experience that had been acquired during the construction of a range of racing yachts and commercial fishing vessels in particular stood them in good stead. Now, with increasing demand, new designs for offshore work could be considered, and existing smacks that had previously worked in

Robert Aldous, shipyard proprietor, on his daily inspection, 1886.

Above: The jetty and No. 1 slipway in Aldous's yard.

Right: A chart published in Aldous's catalogue showing the location of the yard.

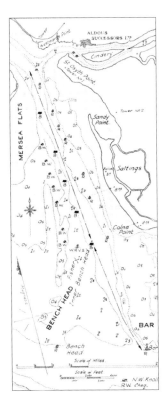

the estuaries or in coastal waters could be extended, upgraded and converted for the more arduous offshore trade. During the 1860s the yard's reputation for strong, fast and reliable vessels attracted commissions for racing and pleasure craft, a demand which they were happy to meet with innovative designs. This complemented the building and maintenance of commercial fishing vessels and cruisers as well as the repair and fitting out work that was vital to the business as the century wore on.

Along with the growth of the oyster fleet and the increased workload for the shipyards came the need for additional maritime trades. As details from the census returns have shown, these ancillary trades increased in number and diversity. Tables 1 and 2 show this growth in detail by comparing the censuses of 1841 and 1881:

Table 1. Numbers engaged in maritime trades in Brightlingsea, 1841.

Apprentice	51
Coastguard	3
Fish Wife	1
Fisherman	29
Mariner	207
Oyster Dredger	2
Oyster Merchant	3
Sail Maker	4
Ship Owner / Captain	2
Shipwright	4
Tide Waiter / Surveyor	4
Total	**310**

Table 2. Numbers engaged in maritime trades in Brightlingsea, 1881.

Apprentice	6	Oyster Merchant	25
Bargeman	2	Rope Maker	1
Chandler	2	Sail Maker	8
Coastguard	8	Ship Owner / Captain	1
Customs Officer	4	Shipwright	50
Ferryman	1	Smack Owner	4
Fisherman	53	Water Bailiff	1
Mariner	315	Waterman	3
Net Maker	1	Yacht Captain	8
Oyster Dredger	2	Yacht Steward	3
Oyster Foreman	1	**Total**	**499**

The launching of the *Millie Colchester*, one of the first vessels built by Stone's yard, 1892.

In forty years the number of maritime occupations and trades rose from eleven to twenty-one. Four categories showed particularly strong growth: fishermen, mariners, oyster merchants and shipwrights. The latter increased by 1,150 per cent, clearly as a result of the increasing demand in the shipyards, although, strangely, the number of apprentices seems to have reduced.

As the nineteenth century wore on, Brightlingsea's shipyards continued to enhance their reputation for excellence. They had competition, of course, particularly from yards at Wivenhoe and Rowhedge, but Brightlingsea seemed to be the hub of the oyster trade. The Aldous yard in particular knew *how* to build oyster smacks, and as demand for the larger, first-class smacks developed they not only built these from the keel up but also converted smaller smacks for the offshore trade (as stated earlier) by lengthening, strengthening, re-rigging and inserting wet wells.[38] Aldous understood that reliable vessels that were fast under sail meant more profit for the owner, skipper and crew, and the local oystermen knew that an Aldous smack would give them exactly that. Other yards such as James' (to become the large James and Stone Shipyard of the twentieth century) and specialist blacksmiths, chandlers and sail-lofts all developed along and near the waterfront as the offshore trade flourished. Rope Walk, a long, straight promenade at the eastern end of the waterfront, was undoubtedly the area where the rope-maker worked. The 1/2500 Ordnance Survey map of 1874 shows

Mudberths, Aldous's yard.

Repairs underway at Aldous.

some forty-four 'out-buildings' on the waterfront surrounding a crane, smithy and the two main shipyards.[39] This 'establishing' of onshore maritime trades as a result of the busy inshore and offshore industries is ultimately confirmed by a series of plans submitted to the new Brightlingsea Urban District Council up to 1912 by Aldous Ltd for permission to build various new workshops, stores and showrooms.[40]

In 1872 Brightlingsea saw the arrival of the Revd Pertwee, following the death of the previous incumbent the Revd William Latten.[41] Perhaps the townsfolk initially had no great expectations of their new vicar, but Arthur Pertwee left his mark on the community in many ways and his name is still spoken with affection in the town today. Shortly after his arrival the *Brightlingsea Parish Magazine* was introduced, to which he was an avid contributor. This was at the time when the offshore oyster trade was approaching its busiest years and first-class smacks were heading for the Terschelling Banks off the Dutch coast during the winter months in search of new deep-sea oyster grounds.[42] Although the new smacks were well equipped to deal with heavy seas and long voyages, Terschelling (or 'Skilling' as the area was known locally) would prove to be a terribly hazardous place to dredge for oysters. In the closing decades of the nineteenth century, *BPM* columns were increasingly filled with the news of Brightlingsea smacks suffering disaster at Terschelling, or on the journey to and from it. The winter gales and hazardous working conditions proved too much for some of the vessels and their crews. In January 1884 Pertwee wrote:

> During the past year five smacks hailing from this port have brought up at the bottom of the North Sea, and in that vast cemetery we have buried twice as many men as in the old churchyard at home! To fight the winter storms of the North Sea on their own ground is an unequal struggle, which bravely as it has been maintained by the mariners of Brightlingsea, it is madness to prolong.

He then went on to list the twelve crew of the smacks *Pride* and *Walter and Henry* (from Dover) who perished a month earlier.[43] In 1891, preaching to a crowded church, he noted that in his eighteen years at Brightlingsea 101 lives had been lost at sea.[44] Pertwee clearly had a great affection and concern for his seafaring parishioners. He often went to sea with the crews to experience something of their working lives, and during stormy weather it is claimed that he would spend the night on top of the lofty tower of All Saints church with a lantern to act as a guiding light to the smacks out to sea making for home.[45] He also instigated a frieze of memorial tiles set around the inside of All Saints church, each tile commemorating the life of a Brightlingsea mariner lost at sea.

Brightlingsea's oyster crews, as experienced seamen, would naturally have had alternative activities to supplement their income. Crewing the new large racing yachts during the summer months for wealthy owners, scallop dredging (a trade that the

Above: Commemorative tiles, All Saints Church.

Left: Canon Arthur Pertwee, M.A.

deep-sea smacks could easily turn to), salvaging and stow-boating for sprats were always good alternatives. Line fishing for cod was attempted but with limited success.[46] The pages of *BPM* reveal that towards the end of the nineteenth century and into the early years of the twentieth, Brightlingsea had indeed became a venue for large, ocean-going racing and pleasure yachts. These vessels would require experienced crews who knew how to handle large vessels under sail in all weathers, and the town's mariners with their offshore experience were clearly well suited to this work. In addition, the town's shipyards were well accustomed to working on racing and pleasure yachts, the Aldous shipyard having designed and built several smaller examples prior to the boom in smack building.[47] Hence, the racing yachts in particular found able crews and favourable facilities in Brightlingsea:

> The yacht of the season [1884] has been the *Lorna*, and her success does honour to Brightlingsea in more ways than one, for not only is she manned by a Brightlingsea crew, but her marked increase of speed this year is the undoubted result of judicious alterations and improvements recently undergone in the shipyard of Messrs. Aldous & Co.[48]

In the closing years of the nineteenth century, despite governmental activity geared to the protection of oyster grounds,[49] the industry went into decline as demand for oysters seemed to be waning. A contemporary report claimed that 'the destruction of these once flourishing industries has been chiefly caused by over-dredging and the want of a proper annual close season.'[50] The winter of 1894–95 brought bad weather

Above: Looking from the causeway toward Aldous's yard.

Right: Syndicate yacht berths.

Above and below: Smacks, dinghies and tenders, 1906.

that hampered offshore dredging and coincided with a dramatic fall in the price of the catch. This came about largely as a result of the linking of the consumption of oysters to recent outbreaks of typhoid fever.[51] For the rest of the century the demand for oysters continued to decline, and Brightlingsea mariners became more engaged in stow-boating for sprats and dredging scallops in the winter months. During the summer months the port was increasingly dominated by the large pleasure and racing yachts which were now a familiar sight.[52]

Crewing yachts seemed to take precedence for the mariners of Brightlingsea in the early 1900s and, whilst sprat fishing was still a winter option, oyster consumption was dealt another blow by the news of a merchant from Emsworth in Hampshire being prosecuted by the local district council. An outbreak of typhoid fever occurred in Southampton, Winchester and Portsmouth, and the merchant was blamed for allowing sewage to contaminate his oysters.[53] In April 1905 *BPM* reported: 'The effects of the depression in the oyster trade are still badly felt. We understand that there is very little prospect of importations this season, the old stocks being mostly still on hand.'[54] By 1910, the year that Robert Aldous died (proprietor of the town's famed shipyard), with more mariners seeking work on large luxury yachts and in the more reliable trade of fishing for sprats (mainly due to new improved methods for curing), oyster dredging was in terminal decline.[55]

The site of the former Aldous yard, 2009.

The harbour area in 2009.

So it seems that the Brightlingsea mariners who remained in the town did keep busy crewing pleasure and racing yachts, stow-boating and scallop dredging up to 1914. But the heady days of the mid- to late nineteenth-century offshore oyster dredging trade, which Hervey Benham described as 'the hardest and cruellest trade Essex men ever took part in', was now firmly in the past.[56] In September 1914, as war loomed, *BPM*, in sombre mood, lamented the cancellation of the town's summer regatta and the uncertainty now surrounding its once-dominant fishing industry.[57]

CHAPTER THREE

An Urban Community in the Making

It has now been shown that Brightlingsea underwent a process of considerable expansion and change during the nineteenth century, developing from a small fishing hamlet to a vibrant, successful and diverse town by the early twentieth. By examining its institutions, both old and new, this chapter seeks to analyse and understand the social and political consequences of this urbanisation, and the effect upon the communal, secular and spiritual life of the community. The impact of growth and economic development upon the population's quality of life as new opportunities and facilities were provided will also be explored.

The manors of Brightlingsea and Moverons were bought by the Magens family in 1763,[58] and the Manor of Brightlingsea held its courts regularly almost throughout the nineteenth century. In the early decades they dealt solely with estate lands. This is typified by the General Court Baron of the Manor of Brightlingsea held on 30 May 1826 on behalf of Magens Dorien Magens, Lord of the Manor, represented by Charles Round, his steward. The business concerned the sale, transfer and surrender of copyhold lands and messuages such as the sale to James Barker, a surgeon from Colchester, of a piece of land in an area known as Orchard Field for £12 10s. Rents were also reviewed at this court.[59] These courts continued in a similar manner throughout the 1830s and '40s, but by the 1860s, with James Robinson of Pontefract, Yorkshire, and Edward Westwood of Old Swinford, Worcestershire, now jointly Lords of the Manor, the courts took on a different character. The biggest change was that court was now held away from the parish. On 22 June 1864 the court was held before George Bradley, steward, at Castleford, Yorkshire, over 200 miles to the north. The business was, however, much the same as before, dealing with land in Brightlingsea – although by now this appears to have been less extensive as more land had been sold from the estate.[60]

By 1869 courts were once again held at Brightlingsea, now at the parish church. The court of 28 October 1869 shows a much more formal and traditional meeting with the View of Frankpledge and General Court Baron being held before George Bradley and George Digby, chief stewards. No less than fourteen local men were sworn in at the Inquisition of the Leet (mostly from the maritime trades) and the common fine of 6s 8d was paid to the Lords of the Manor according to ancient custom. Bread and ale tasters were chosen, and after the main business of the court was concluded (again solely concerned with the administration of manorial land) the court adjourned to the Swan Inn.[61]

One of the later set of minutes available for the Manor of Brightlingsea, from 22 October 1879, again shows considerable emphasis on tradition and ceremony. The View of Frankpledge was held on behalf of Edward Westwood and George Bradley, now Lords of the Manor, before George Bradley (junior), deputy steward. As before, the Leet jury consisted mainly of names from the marine trades: R. Francis, R. Aldous, A. Aldous, J. Aldous, W. Minter, G. Peggs, C. Blyth, J. Folkard, E. Aldous, J. Salmon, E. Ainger and G. Coppin.[62] Bread and ale tasters were chosen, along with water servers, and G. Ruffel was elected reeve and bailiff. Homage was sworn by R. Francis and R. Aldous in the Court Baron. The court, however, now dealt with considerably less business concerning manorial lands, although several surrenders are recorded. Thus, the later courts are ostensibly ceremonial affairs dealing with a much reduced workload.[63]

Brightlingsea's manorial courts *could* have gradually declined in their status during the nineteenth century, as many others throughout the land did.[64] But this was certainly not the case – although the amount of business conducted by the courts fell, particularly after the land sales during the building phase from 1860 onward, they in fact appear to take on a far more traditional and ceremonial feel, and have higher attendances. This coincides with Brightlingsea's growing urban status and sense of communal self-identity and importance, and must surely be seen as a symptom of this. The preservation and development of the manorial courts with the support of many from the marine trades suggests that the emerging lively town wanted to hold on to and cherish an aspect of its historic past amongst its existing and emerging institutions, and, indeed, to use and adapt them to new circumstances.

Brightlingsea was an ancient parish in the county of Essex. From 1809 the Revd John Robertson, MA, oversaw the select vestry, which met regularly to adminster to the parish needy. At the meeting of 19 April 1829 Robertson appointed Thomas Syer and Robert Mason as churchwardens, and the business of dealing with requests for financial and practical assistance was attended to. Many requests for financial help were turned down, but some were granted: Mr Ward received 3s as he was unwell, and Mrs Gild's request to have the doctor in her confinement was granted.[65] During the early decades of the nineteenth century many requests for assistance were made

Above: All Saints church, early 1900s.

Right: All Saints church, 2009.

to the vestry (around ten to twenty each month), mainly from families in the marine trades.[66] The vestry rate book for 1830 lists some forty-four occupiers of oyster layings and ninety vessels which appear to be oyster smacks. Many indentures were also made at this time by the churchwardens and overseers of the poor placing young men into apprenticeships with local oyster dredgers and fishermen.[67] Here, once again, we see an ancient governmental institution being adapted to new circumstances.

A highly significant development connected with the urbanisation and growing ambition of Brightlingsea was the consecration in 1836 of a new Anglican chapel-of-ease, Saint James'. This sizeable brick-built chapel in the High Street saved many

St James's, the Anglican chapel-of-ease.

Details from the poor relief book of 1827–28, including some imaginative doodling!

from the mile-and-a-half journey out of town each Sunday morning to the isolated medieval parish church.[68] Parish minute books, after the appointment of the Revd William Latten in the same year, show that the vestry continued to meet at the parish church until 1862 when a meeting was called 'for the purpose of making an appeal to the Poor Law Board for the removal of the vestry from the Parish Church, [and] that a room may be provided in the [High] street for all such said vestries to be holden for the accommodation of the rate payers.' The following year meetings were held at the National school, but with a seemingly reduced workload.[69]

In 1872, prior to the appointment of Canon Pertwee, a sequestration order was made on behalf of the Bishop of Rochester appointing Edward Stammers and Robert Folkard (churchwardens) as 'sequesters and special collectors for the Parish of all fruits, tithes, rents, profits and other ecclesiastical rights and emoluments, upon the death of Revd William Latten.'[70] Although this shows that the vestry was still clearly active in its collections at this stage (prior to the Local Government Act of 1888), there is less evidence of payments to the poor. In the absence of surviving minute books it would be unwise to speculate too much, but it does seem as though there was less demand upon the vestry for relief in the later decades of the nineteenth century. This in turn suggests that the community had become generally more affluent and less reliant (particularly in the marine trades) then it appeared to be in earlier decades. Clearly the community still had residents who were in poverty or infirm and unable to work, but records suggest fewer payments were made to such people as the century wore on.[71]

Canon Pertwee arrived in the town at a time when the maritime trades were approaching their most lucrative yet demanding and treacherous period.[72] Dr Dickin remembered Pertwee as a man who had a special allegiance to seafaring families: as mentioned in chapter two, he often went to sea with the oyster crews and showed great courage when nursing the crew of a ship lying off Brightlingsea stricken with smallpox. Influenced by the Oxford Movement, he introduced a more ornate ritual to services, and was much involved in local affairs.[73] When *BPM* first appeared in 1882, Pertwee, as has been shown, was an avid contributor. His reports on the fortunes of the oyster and sprat boats are a valuable source of historical information. He must also be remembered for the installation within All Saints church of the frieze of commemorative tiles dedicated to the memories of Brightlingsea mariners lost at sea.

The description of Brightlingsea given by Major Reynolds in 1797 (chapter one) allows us some insight into how the community regulated and managed its local affairs at that time. It informs us that the inhabitants had 'formed some regulations among themselves', and that 'Mr. Tabor, a principle proprietor in the oyster trade, acts as Mayor'. Reynolds' use of the word 'mayor' should not be taken literally, but it does indicate that the concept of a leading office was still around. In fact, Tabor is thought to have been the Cinque Port Deputy, although the title was falling into disuse by the late eighteenth century.[74] With the appointment of Commisioners to organise

Far left: The entrance, All Saints church, 2009.

Left: John Bateman, M.A.

the Poor Law Unions in 1834, Brightlingsea's administrative structure was brought into line two years later when it became part of the Tendring Hundred, Lexden and Winstree Poor Law Union. At this stage, however, there is no evidence of any real 'urban identity' within the community. After the Public Health Acts of 1873 and 1875 created new authorities with responsibilities toward public health, rural areas were divided into sanitary districts to be co-terminous with the Poor Law Unions, and in 1880 the town (as it could now be considered) became part of the Tendring Rural Sanitary District.[75]

Although still classified as part of a rural district, a sufficient collective feeling of urban awareness and municipal aspiration had now developed for local dignitary John Bateman, Canon Pertwee and others to revive the old connection with Sandwich in Kent. From 1885 Brightlingsea unilaterally re-established its status as a non-corporate member of the Cinque Ports. A deputy was once again elected by the freemen of the town, and presented with John Bateman's gift of the 'Great Opal' and silver chain of office. This was to be worn by the deputies until such time as they became mayors in their own right. [76] The mount of the opal bears the inscription *Urbs Brictriceseiae ex dono Johannis Bateman* (The Town of Brightlingsea by the gift of John Bateman), and *Pulchra Matre Filia Pulchrior* (From a beautiful Mother, a more beautiful Daughter). The solid silver chain is made up of alternate links of crossed sprats and oyster shells, reflecting the industry upon which the town flourished. [77] The 'Great Opal' and chain are still worn by deputies to this day.

Further administrative changes after the Local Government Act of 1888 saw the town (now with its own parish council) become part of the Tendring Rural District in 1894. However, the urban status and identity that was clearly felt by the townspeople with their revived Cinque Port Liberty status could no longer be ignored and was recognised two years later when the Brightlingsea Urban District was created.[78] The population of the town, as has been discussed, had grown considerably whilst that of the surrounding district had shrunk or stagnated. Indeed, it is claimed that amidst this new urban awareness Brightlingsea's ultimate goal was to become a borough.[79] Within a few decades, however, it would become clear that this would never be realised.

Brightlingsea's sense of urban identity and municipal aspiration were, although probably genuinely felt throughout the town, enhanced and promoted by key individuals of standing. Some of those who made significant contributions were not originally from the locality: Mr John Bateman, MA (Trinity, Cambridge) purchased Moverons, the Rectory Hall and Lodge Farm, in 1871. He is described by Dr Dickin as Deputy Lieutenant for Staffordshire and JP for Essex, in addition to representing Brightlingsea division on the Essex County Council and serving as a member of the Brightlingsea Urban District Council and the Kent and Essex Fisheries Committee.[80] Canon Arthur Pertwee, after gaining his MA from Pembroke College, Oxford in 1863, served as curate at Brancepeth, County Durham, until 1864, and was then vicar of St Margaret's, Leicester, before moving to Brightlingsea in 1872.[81]

In the nineteenth century Brightlingsea acquired a permanent police presence. The census of 1851 shows one police officer and one customs officer resident in the

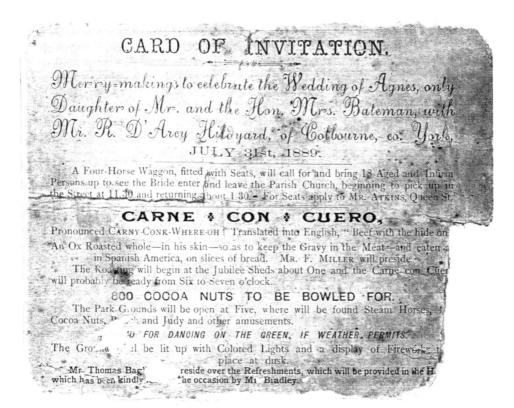

An invitation to 'the merry-makings to celebrate the wedding of Agnes, only Daughter of Mr. And the Hon. Mrs. Bateman', 31 July 1889.

town.[82] After 1800 recorded crime seems to have been varied in nature: accounts of prisoners from Brightlingsea at the Essex convict gaol are sparse, but do show that John Clark and Thomas Lownes were convicted in 1829 for not maintaining their families. In 1826 three Brightlingsea men were convicted for assault. In 1829 John Bird was convicted for 'uttering counterfeit money', and in 1833 two more men were convicted twice for assault, whilst in the same year E. Dennis was convicted for poaching.[83] Crime rates in the surrounding rural areas were generally low, but with a rapidly expanding population and the growing importance of Brightlingsea as a centre of the oyster trade the police presence in the town was increased to two in 1881 and five by 1901.[84] In fact, a permanent presence was maintained essentially to police the oyster trade: disputes on and off the water were not uncommon, and in 1891 the Colne River Police were formed as part of the Colchester Borough

Charles Christmas Trubshoe, Brightlingsea's first policeman. He retired in 1884, being replaced by Sergeant Arthur Brown and three constables.

Police to protect the lucrative inshore oyster fishery in the river, down to the estuary at Brightlingsea.[85]

During the eighteenth century dissenting congregations in England and Wales had dwindled considerably, but after the 'Evangelical Revival' around the turn of the century numbers rose again dramatically. In 1851 the national census of religious attendance revealed that the various nonconformist denominations accounted for nearly half the worshipping population. Also, it was shown that more than half the population did not attend any form of service, and in the new industrial cities the proportion of attendances was as low as one in ten.[86] Dissent in the rural areas of Essex during the nineteenth century was not uncommon, but Alan Everitt found that it particularly flourished in the 'industrial villages' or craft centres in East Anglia and the Midlands. In addition he discovered that in these parishes, where essentially proletarian sects such as the Primitive Methodists might be expected to dominate, this was not the case. The Independents or Wesleyans who, he suggests, 'usually comprised a rather more prosperous element in the local community in these counties' were far more numerous. This would suggest not so much the expansion of the working class in these areas but the emergence of a lower middle class.[87]

Brightlingsea had a growing and diverse nonconformist community during the nineteenth century. The Wesleyans are said to have arrived in 1796, meeting at various private venues until the first wooden chapel was built in 1804 to the north of 'Hearse Green'. Steadily increasing attendances are reported, and the building was extended in 1822.[88] By 1842 the enlarged chapel is said to have been far too small and overcrowded, and, despite the congregation being generally poor, a new larger building was planned.[89] In 1843 members numbered 103, with attendances of up to 300, and plans for a Sabbath School were put forward. The foundation stone of the new Wesleyan chapel was laid on this site on 18 July 1843, and in 1861 plans were drawn up for a Wesleyan school.[90]

In 1809 Dr Moses Fletcher is credited with introducing the New Jerusalem church, based on the doctrines of Emanuel Swedenborg. A congregation was formally founded upon the ordination in 1813 of Mr Arthur Munson as the New Church minister for the town. Despite a shortage of funds, a chapel was built in 1814 at a site that was to become 18 New Street.[91] The Essex Quarter Sessions received a letter dated 15 September 1829 giving numbers at the New Jerusalem congregation at about 150, and thereafter numbers continued to grow steadily.[92] However, relations between the New Church and the Wesleyans during the early nineteenth century were poor and caused social division and unrest in the small community. With growing attendance and support, a much larger and imposing New Jerusalem chapel was built in 1867 at the north-west end of the High Street.[93] Relations between the Wesleyans and the New Church in the second half of the nineteenth century seem to have improved. The numbers of both sects expanded as members joined from farther afield, and a more tolerant, even

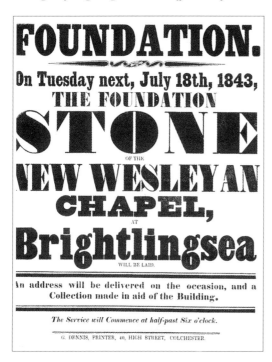

Poster advertising the laying of a foundation stone for the Wesleyan chapel, 1843.

The Methodist chapel, 2009.

co-operative, climate emerged. Data from the examination of registers indicates that membership of the New Church was generally from poorer groups, which included oyster dredgers and sail-makers.[94]

A letter received at the Quarter Sessions in January 1841 requested that a building in Brightlingsea belonging to George Johnson be certified as a chapel for 'Protestants dissenting from the Church of England.'[95] Unfortunately it is unclear which sect this letter refers to. The Congregational church did have a wooden chapel on John Street built around 1830. By 1846 attendances grew to around seventy-five or more (many of whom were mariners),[96] so more land, this time near Hog Lane, was purchased to

The New
Jerusalem chapel.

Letter, dated
1829, giving
details of the
New Jerusalem
congregation.

house a new, larger chapel. This was completed in 1864,[97] and by 1889 correspondence suggests a lively, well-attended church.[98] A register shows 116 members in 1891 and 147 in 1904, many living in the new terraced housing leading down to the waterfront.[99] The Primitive Methodists were one of the last nonconformist groups to establish a chapel in the town, firstly at New Street, followed by a larger building known as the Coronation church built on Ladysmith Avenue in 1902.[100] The Salvation Army headquarters was first built at Hurst Green before moving to a new, grander building in Tower Street around 1908.[101]

Within the town's nonconformist community a pattern of substantial growth and development can, therefore, be observed. Small congregations meeting in houses or small makeshift chapels developed throughout the nineteenth century, and later acquired new and larger premises. These were constructed on land made readily available and were financed by public subscription, funds from the wider organisation of the various sects, and bequests. Although nonconformist congregations existed in the town prior to the mid-nineteenth century, cross-referencing of census surnames and chapel records strongly suggests that their numbers were swelled by those moving to the town and working in the marine trades.

Letter, dated 30 January 1889, inviting a friend to a 'sixpenny congregational tea'.

The former
Congregational
chapel, 2009.

From the early nineteenth century Brightlingsea had basic facilities for schooling.
A report published in 1819 describes a school (later to become the National school)
'containing 12 children; the master of whom has £10 per annum, with a house and
garden; the funds are the interest of £130. 3 per cents, and the rent of an oyster-laying.'
A Sunday school and four day-schools are also mentioned, and this report concludes
with the observation that 'the poor [in Brightlingsea] have the means of education,
but appear very indifferent in taking advantage of them.'[102] This seems reasonable
for a rural community of little over 1,000 inhabitants in the early nineteenth century,
mostly engaged in the oyster fishery. The Rural Dean's return for 1844 gives the
infant school connected with the parish church as having forty-four boys and forty
girls while the Sunday school had forty-six boys and sixty-five girls.[103] In 1842 the
National school gained a parliamentary grant of £88, and in 1861 the new Wesleyan
school received a massive £438 from the same source.[104] With most chapels having a
Sunday school (in addition to the Anglican one) and an increasingly well-attended
National school, Brightlingsea began to achieve an enviable record in education,
in contrast to the reported situation during the early 1800s. The Wesleyan school is
reported to have had twenty of its twenty-four children taught drawing, and when
examined at first and second grade three pupils excelled and obtained prizes. This
school's boys are also singled out as having a high standard of arithmetic.[105] The
numbers taught drawing at the National school, Station Road, are given as 170 for
1878. For the same year and subject the Wesleyan school had 151. Indeed, in his
report to the Lords of the Committee of Council on Education in 1882, H.B. Rowan
lists Brightlingsea National (Girls) and Wesleyan schools as amongst the best in the
district. [106]

Station Road and 'The Palace', early twentieth century.

The provision of education at Brightlingsea clearly improved through the nineteenth century with the growth of both the National and Wesleyan schools supplemented by Sunday schools of various denominations. A more affluent and socially aware community was developing; one that looked to education as an important opportunity for the young, and had the means to support this on an individual basis, within the community. In Brightlingsea, at least, many by the end of the nineteenth century had the advantage of an improved standard of educational provision.

The town was fortunate enough to have had resident medical doctors or surgeons for many years. J.C. Parker, formerly of the Royal West Middlesex Militia, was in practice prior to 1815: he also became medical officer of the Lexden and Winstree Union. His son, George Parker, MRCS, practised from 1852 until 1898, and William Bird Parker, MRCS, was resident from 1849 until moving to St Osyth in 1857. Charles Ling is shown as practising in 1892. He was the district medical officer, public vaccination officer for the Tendring Union, the Admiralty agent and surgeon, and medical officer of health for Colchester Port.[107]

During the nineteenth century, in addition to gaining new opportunities for religious worship and education, Brightlingsea saw the quality of its communal life enhanced by the emergence of several philanthropic and social institutions, both religious and secular. The *Essex Standard* for 8 February 1860 reported 150 persons meeting at the Temperance Hall, Brightlingsea, for the bi-annual meeting of the Working Man's Improvement Society, over which the independent minister, the Revd E. Pay, presided.[108] A Masonic lodge, Lodge of Hope (No.433), formerly

The Masonic temple, 2009.

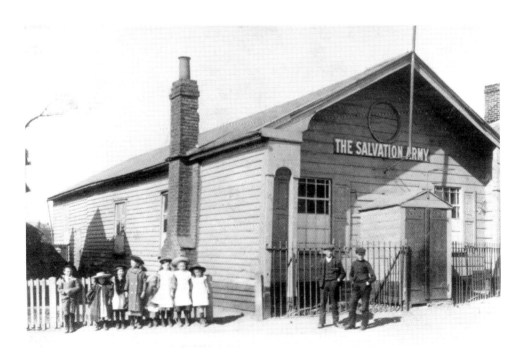

The Salvation Army hall.

The Salvation Army hall, 2009.

a maritime lodge meeting onboard a ship, met at various public houses in the town from the mid-1800s. In 1907 a new Masonic temple was consecrated by the provincial grand master just yards from the site of the new Salvation Army hall.[109] Also in this area, close to the centre of the town, the Ancient Order of Foresters had a meeting house, 'Lifeboat' Court, No.4390. The Foresters had existed since 1864 when they were formed from two earlier friendly societies.[110] This rather grand hall has been the venue for many local functions and meetings, not least for the parish and urban district councils. A new YMCA building appeared at the eastern end of the High Street around the turn of the century. Sports were organised and played regularly in the town, and the Brightlingsea cricketers were reported to have won every match that they played up to August 1884.[111] However, it was not until the twentieth century that a football club was formally established.[112] *BPM* reports on a gift being made by the reclusive millionaire Mr Bayard Brown of the S.Y. *Valfreyia* of the recreation ground in Regent Road which was opened in 1893 with a pavilion being built as part of Queen Victoria's diamond jubilee celebrations. See Appendix Two for further information regarding Mr Bayard Brown's association with Brightlingsea.

This survey of Brightlingsea's institutions during the nineteenth century does support the view of a community undergoing rapid transformation. The changes that take place are symptomatic of, supported and driven by, a community that is growing

A rare photograph of
the reclusive millionaire
Bayard Brown.

Sports day on the recreation ground, 1904.

A procession through the town, Coronation Day, 1902.

Coronation Day, 1902.

Silcott Street: the Co-Op, founded 1912.

The Railway Tavern, 2009.

in size and is becoming more affluent and 'urbanised'. This growth, fuelled by the maritime trade, presents the opportunity for the town to enhance and confirm its urban status, which is exemplified above all by the re-establishment of the link with the Cinque Ports and the creation of an urban district. The rise of philanthropical and social institutions so typical of larger Victorian towns and suburbs, diverse and growing nonconformist congregations relying heavily on the maritime workers and their families, the growth and achievement of educational facilities coupled with diminishing evidence of poverty and increasing opportunities, all point towards a community living in, and benefiting from, a period of highly localised economic and social expansion.

CHAPTER FOUR

Population History

Brightlingsea's decennial population statistics from the nineteenth century suggest that it was a period of almost continual growth (Fig. 1, Chapter One). This chapter will look at this period in more detail and make comparisons with the national picture of population growth at this time, as well as with key fishing centres and other local communities. This will also include an examination into the origin and extent of migration, occupation and gender profiles, the migrational 'pull factor' generated within the town, and the 'push factor' in the agricultural hinterland and more distant fishing communities.

In order to achieve these aims data has been extracted from the census returns for Brightlingsea for the years 1841, 1851, 1881 and 1901. There have been several approaches suggested for analysis of such data;[113] in this case the initials, surnames, gender, occupations, ages and birthplaces of all persons engaged in marine activities were taken from enumerators' books and entered into a purpose-built database. In addition, details of persons engaged in non-maritime trades or occupations were similarly entered. More data was gathered from printed census abstracts for neighbouring towns and villages such as Wivenhoe, Thorpe-le-Soken, Great Bentley and Saint Osyth, and for comparable fishing communities further afield. [114]

But clearly the use of census data has limitations. Therefore, one must review the way in which census returns were originally created, and remain aware of their inherent shortcomings. With no inter-censal data, as E. Higgs observes, the returns are just a sample or 'snap-shot of society on one night every ten years'.[115] Other limitations are well documented, including the omission in 1841 of the 'relation to head of family' column.[116] As that census also gives only the county of birth rather than the actual parish or town, this makes the analysis of localised migration less effective prior to 1851 when the parish or town of birth was first included. An awareness of total migrational flow is equally important;[117] census data showing that Brightlingsea saw a net population growth during the nineteenth century, but the number of those who left the town is far less clear.

This quantitative method does have advantages. Brightlingsea was a relatively small community, and it has therefore been possible to extract all the entries in the enumerators' books for individuals in maritime occupations for each of the selected years, rather than samples of the data (say, one in every five households), as some analysts suggest for more wide-ranging studies.[118] Using the original enumerators' books (on microfilm) also allows a greater range of detail from individual entries to be included in the data set.[119] In addition, parish boundaries have remained constant throughout the period in question, and therefore no adjustment for this is required.[120] The inclusion of birthplaces after 1851 allows a spatial analysis as these can be mapped to show the range of migrational in-flow.

Between the years 1841 and 1901 England's population roughly doubled. Fig.2 details this growth, which at the time was largely an urban phenomenon.[121] With increasing transport links and the labour demands of urban industrialisation, rural to urban migration often left rural areas with either declining or, at best, static populations. Brightlingsea experienced similar growth during this period, slightly exceeding the national rate. This, however, was not the case for other rural parishes and agricultural communities in the region. Tendring, the Rural District and Hundred surrounding Brightlingsea saw a net population growth throughout the nineteenth century, but, again, census statistics show this to be fragmented, centred on certain areas.

The High Street, May 1903.

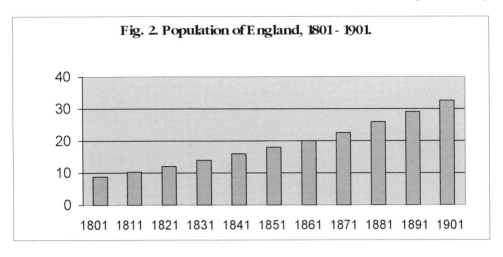

Fig. 2. Population of England, 1801 - 1901.

Tendring District was essentially agricultural, with the exception of the ports of Harwich and Mistley in the north and more latterly the developing holiday and excursion centres of Clacton, Frinton and Walton on the coast. Colchester, the largest town in the region, situated just outside the administrative boundary of Tendring, saw increasing industrialisation and growth during this period.[122] However, the population growth that occurred at Brightlingsea was not seen in neighbouring (mainly agricultural) communities such as the adjacent coastal settlement of Saint Osyth, the inland communities of Thorpe-le-Soken and Great Bentley, or even Wivenhoe, a neighbouring town whose fortunes were also linked to the fishing and oyster industry. Fig. 3 illustrates this in more detail. The figures for Wivenhoe after 1881 are not included as the subsequent boundary changes introduce a misleading skew on the data.[123]

The enumerators' books show that the majority of the working populations of Saint Osyth, Great Bentley and Thorpe-le-Soken were engaged in agriculture. It is not surprising then that many would seek opportunities in nearby Brightlingsea that did not exist in their own communities.

The census for 1841 shows James Herbert (twenty-two), a labourer, and his wife Sarah (twenty-one) with their children Charlotte (fifteen), William (twelve), Jacob (ten) and Susan (eight) living at 93 Plough Road, Great Bentley. Ten years later all these children appear to have left home and three more to have been born. By 1881 William Herbert, now a mariner aged fifty years (according to the enumerators' book), moved to Brightlingsea. He and his wife Charlotte (from the town) had six children: William and Albert (also mariners); Edgar was a baker; Jessie, Arthur and Katie were at school. William's extended family was still in Great Bentley, being mostly agricultural labourers. In 1861 George Wringe (eleven) lived with his parents William (an agricultural labourer) and Sarah at 87 Angers Green, Great Bentley. By 1881 George, now a mariner aged thirty-two, had moved to Brightlingsea and married

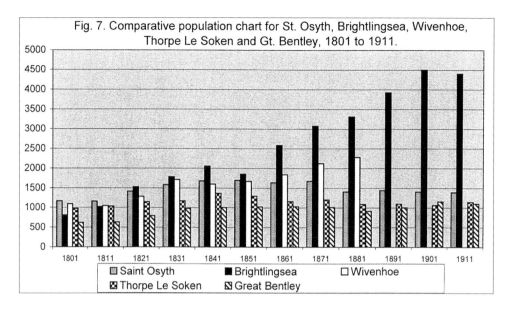

Fig. 7. Comparative population chart for St. Osyth, Brightlingsea, Wivenhoe, Thorpe Le Soken and Gt. Bentley, 1801 to 1911.

Tamara (from the town), and they lived at 4 Sidney Street with four children. Most of George's extended family remained in Great Bentley, his male cousins and nephews working as agricultural labourers. Mariner Joseph Taylor (thirty-eight) and his wife Maria lived at 49 Sidney Street, Brightlingsea, in 1881, but Joseph had been born into a large family of agricultural workers in Great Bentley. In 1861 Samuel Austin (five) lived at his birthplace, Saint Osyth, with his parents Samuel (an agricultural labourer) and Martha. Also at home were his two elder brothers William and Henry, both agricultural labourers. By the age of twenty-five Samuel, now a ship's carpenter, lived with his widowed mother and her granddaughter (who worked as a charwoman) in Brightlingsea.[124] These are just a few examples of individuals who moved from the local agricultural communities to take advantage of opportunities in the nearby port. The statistical details of this local migration to Brightlingsea are examined in greater depth later in this chapter.

Fig.4 shows population details for Brightlingsea and four other coastal fishing ports which for the 1801 census (with the exception of Wells) had populations of a similar size. Shoreham in Sussex was also chosen for comparison for having a substantial and ancient fishery, good local oyster grounds (which Brightlingsea mariners were accustomed to visit in pursuit of spat[125] for their own layings), and a history of shipbuilding.[126] Its population, however, did not experience the growth seen at Brightlingsea during the nineteenth century. Cromer and Wells on the Norfolk coast also established fishing communities and had a mixed experience of population change. Cromer, described in 1870 as carrying on a considerable fishery in herrings and lobsters, saw little growth until 1881, and as the town also offered good hotels,

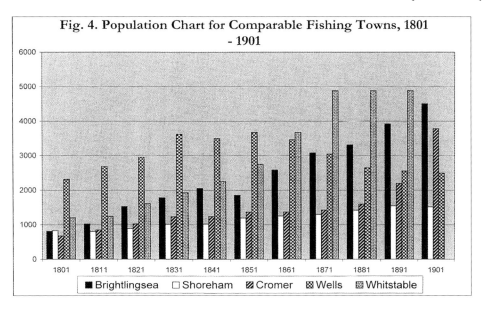

Fig. 4. Population Chart for Comparable Fishing Towns, 1801 - 1901

lodging houses and an excellent beach, this later growth may well be attributable to the development of the tourist trade.[127] Unlike Brightlingsea, the population of Wells, although comparatively large in 1801, actually declined after 1851 despite having a well-established oyster and mussel fishery and a busy commercial wharf.[128] Whitstable had two factors in common with Brightlingsea: a long history of oyster fishing and good railway links to London. The town differed in that it had a substantial coal trade, and was also developing as a resort.[129]

In its population statistics Brightlingsea seems to have more in common with larger more developed ports such as nearby Harwich or Brixham in Devon. By 1860 Harwich was also an established deep-water commercial harbour with facilities for large cargo and fishing vessels. As well as its fishing trade Brixham at this time is described as having had considerable shipbuilding and rope-making businesses. It also had a healthy import and export trade with the Channel Islands and France.[130] Brightlingsea, Brixham, Harwich and Lowestoft all enjoyed good communications with established markets, all of which were enhanced by the coming of the railway. The smaller ports of Cromer and Wells were also linked to the developing rail network, but the effect on their maritime industries does not appear to have been substantial.

The population trends for Brightlingsea and these larger fishing ports (Fig. 5) follow similar growth patterns, with the clear exception of Lowestoft which sees greatly accelerated growth after 1851. In fact, the larger East Anglian fishing ports of Lowestoft and Yarmouth did enjoy a boom in their offshore trawling industries during the latter half of the nineteenth century, and this appears to be reflected in their population figures. However, each town had a quite different experience of this: the Yarmouth

The beach, Brightlingsea.

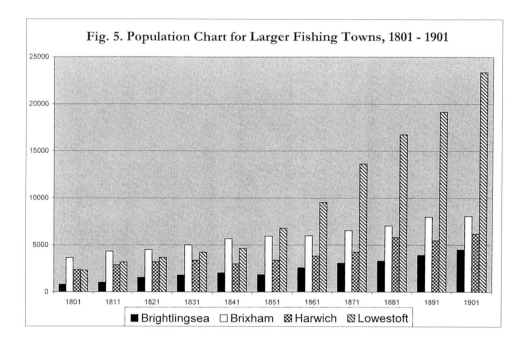

trawling trade went into decline after 1881 to the point of total collapse in 1901, but Lowestoft saw an increasingly prosperous trawling industry develop from virtually nothing from 1860 until well into the twentieth century. This was, however, dwarfed by the expansion in the north of England, particularly at Hull, which was close to the coalfields that fuelled the steam trawling fleets.[131]

All these factors serve to illustrate that during the nineteenth century coastal fishing ports large and small along England's south and east coasts had different experiences, largely dependant on local circumstances, communications and practices. This situation was complicated further by some of these ports simultaneously developing as coastal resorts.

In 1841 the census returns for Brightlingsea reveal a small but established maritime community with a significant proportion of the population engaged in marine occupations. In a total population enumerated at 2,055 people, 310 persons were occupied in marine activities. Only two of these, Anne Baines (a mariner) and E. Ellis (a fish-wife), were women. By 1851 the total population had fallen to 1,852, with 224 persons recorded as occupied in marine trades, Mrs Roots (a smack owner) being the only woman. Eleven different maritime trades are found in the 1841 enumeration, the most significant in terms of numbers being mariners, fishermen and apprentices of no specific trades.[132] Less populous trades included sail-makers, shipwrights, tide waiters,[133] coastguards, oyster merchants, oyster dredgers, ship-owners and a fish-wife. By 1851, although fewer in personnel, thirteen marine trades appear with the addition of barge owners, net-makers, chandlers and smack owners, but without the tide waiters and the fish-wife. The mariners and fishermen again accounted for the largest numbers.

That Victorian census returns regarding occupation are too generalised and lacking in detail is well known, and much work has been carried out attempting to clarify this situation.[134] The term 'mariner' in this study is a good example. We must assume that 'mariners' were occupied in more than one specific activity, while general crew members turned their hands to various activities including the offshore oyster trade.

Barges in the creek.

Cottages near Coastguard Station, early 1900s.

As some of the names listed as mariners are known to be offshore oyster-men, this would seem to back up the assumption that mariners were occupied in more than one specific activity. The inclusion of 'fishermen' in the returns also adds an element of confusion, so we must assign this to the vagaries of the way in which the census data was gathered.

In 1881 a total population of 3,311 in the town included 497 men and two women in marine trades. Mariners and fishermen again formed the bulk of this number, along with an increasing number of shipwrights and oyster merchants. Twenty-one marine trades are given at this enumeration, with customs officers, ferrymen, rope-makers, yacht captains and yacht stewards appearing for the first time. As absentees were now enumerated, 198 men (in addition to the 497 detailed above) are listed as 'away at sea'. By 1901, in a population of 4,501, some 669 men and no women were engaged in twenty recorded maritime occupations. Thirty men and one woman were listed as 'away at sea'.

As can be seen from the details above that Brightlingsea's marine trades were overwhelmingly male occupations. Fig.6 gives us the gender distribution in the town for each of the census enumerations. Curiously, the male population appears slightly higher until 1851, and thereafter the situation is reversed. This again may have much to do with the irregularities of data collection prior to 1851, and possibly the fact that the developing offshore industry initially attracted more single males to the town.

During the nineteenth century the remainder of Brightlingsea's working population were engaged in occupations that have been categorised into thirteen separate groups.

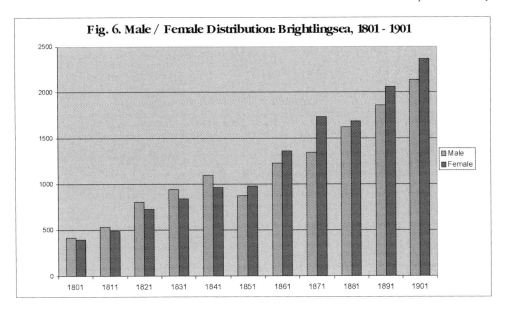

Fig. 6. Male / Female Distribution: Brightlingsea, 1801 - 1901

Agriculture was the largest non-maritime activity, followed by 'other craft' activities (such as dressmakers, tailors, cobblers and other light manufacturers), domestic staff, retailers and latterly building and construction workers.

Table 3 shows the range of occupations along with the numbers and gender of those involved. The vast majority of women who worked in the town at this time were employed as domestic staff, whilst some are shown as living from their own means. The pattern remains similar for the 1841 and 1851 populations, but with higher numbers of women working as professionals such as nurses and schoolmistresses for the latter date. Although the overall population appears to be in decline between 1841 and 1851, the number of agricultural workers increases. Once again it is important to bear in mind the limitations of the pre-1851 census data before drawing too many conclusions from this, for although the 1841 data does allow a useful insight into the situation, a more reliable picture can be constructed after 1851.

Table 3 also shows only a modest overall increase of 58 per cent in the agricultural sector, with a sharper increase in numbers (particularly of women) working in the craft and retail sectors. Craft and retail workers increase by 736 per cent from 1841 to 1901. Numbers of women and girls working as domestic staff rose by 184 per cent. The town's prosperity and growth is also reflected in a 631 per cent increase in the number of construction workers, and a 340 per cent increase in the number of merchants. The arrival of the railway in 1866 provided not only local employment but considerably improved the town's communications with London.[135]

Above: Herbert Ormer's butchers shop and Minter's, around the turn of the century.

Above right: Alice De'Ath's tobacco shop, 1903.

Table 3. Analysis of Non-Maritime Trades: Brightlingsea, 1841 – 1901

Trade	1841		1851		1881		1901	
	Male	Female	Male	Female	Male	Female	Male	Female
Retail / Shop	14	0	26	8	39	14	116	26
Agricultural	91	2	110	1	134	1	147	0
Other Crafts	22	0	26	20	35	82	62	97
Professional	6	0	8	19	18	26	39	40
Domestic Staff	29	50	2	66	3	106	4	142
Ind. Means	3	34	2	2	9	4	26	30
Mill Workers	2	0	3	0	4	0	4	0
Constuction	19	0	16	0	47	0	139	0
Merchants	10	0	6	0	21	0	43	1
Publicans	7	0	7	0	10	2	14	4
Railway	0	0	0	0	9	0	19	0
Clergy	1	0	3	0	5	0	9	1
Police	0	0	1	0	2	0	5	0
Totals	**204**	**86**	**210**	**116**	**336**	**235**	**627**	**341**

By 1901 occupational statistics for those individuals working in non-maritime trades reflect the peak of a period of growth and development which had seen the arrival of the railway, a period of house-building and commercial development, and the expansion of shipbuilding, oyster fishing and the marine trades generally. Table 3 also reveals that agricultural workers had increased in number since 1881, and that more women than men worked in the craft sector. An increase in construction workers, merchants, publicans and railway workers is also apparent. In 1881 the majority of Brightlingsea's working population had been employed in maritime occupations; the remainder of the workforce were, with the possible exception of the agricultural workers, feeling the effects of a healthy local economy fuelled by the town's marine activities. By 1901 the numbers working in other occupations exceeded the marine workforce as the economy and population grew further, in a sense 'catching up' with the marine trade.

The data detailed above starts to support the view of a town growing and developing from the mid-nineteenth century onward with, as Fig.7 illustrates, a constantly high proportion of its working population engaged in marine occupations: of the total workforce at Brightlingsea 52 per cent in 1841 and 41 per cent in 1851 were engaged in maritime trades; the proportion for 1881 peaked at 55 per cent, and in 1901 had dropped back to 42 per cent. In the chart below the three categories of marine, non-marine and others make up the total male and female population figure for each census year. The 'rest' represent the number of non-occupational members of the community such as housewives, children and the infirm. The non-marine category represents those workers and occupations detailed in Table 3.

As discussed, the demography of nineteenth-century England is characterised not only by a significant increase in the population but also by an increasing incidence of migration. English rates of annual net migration are estimated to have increased substantially after 1781 until the mid- to late nineteenth century.[136] For London and the Home Counties[137] between 1851 and 1861 the population showed a net increase of 18 per cent, and 19, 19, 17 and 16 per cent respectively for the decades leading up to 1901.[138] Indeed, nineteenth-century society is described as mobile, 'with largely unrestricted movement, an active land market, a large landless yet rural population, and rising urbanisation'.[139]

The use of censuses, or indeed any other available sources for the period, cannot, however sophisticated the methodology employed, give a truly accurate assessment of the full nature of migration in England during the nineteenth century, not least because only birthplaces are given, and clearly this can only partially reflect the full extent of migrational movements.[140] Nevertheless, the birthplace data from Brightlingsea's census returns should at least give an insight into inward migrational trends for what is after all a small town.

E.G. Ravenstein delivered a paper in 1885 in which he described a number of 'laws of migration' to explain and predict migrational trends within and between

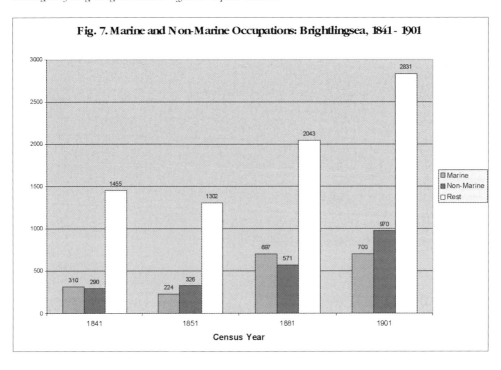

Jacobs Hall and shop, *c.* 1915.

Close view of the shop at Jacobs Hall.

Jacobs Hall, without the shop.

nations.[141] These 'laws' have been variously criticised, praised and refined in the intervening years, but remain generally accepted and valid.[142] In his pioneering work Ravenstein observed that the largest number of migrants travel shorter distances, and towards the larger commercial centres; and that people in urban areas migrate less than those in rural areas. More recently J. Long suggested that 'urban migrants were positively selected – for being the best of the rural labour pool – and that the economic benefits of migration were substantial.'[143] However, a recent study of rural villages in south-east Shropshire using the 1881 census, published in 2006, suggests that the traditional view of workers migrating from the countryside to the large towns and cities was oversimplified. This study, in fact, found that most of their sample did not move to urban locations but to rural ones, utilising their traditional skills.[144] The following analysis of birthplaces in Brightlingsea's census returns of 1881 should indicate the extent to which migrational trends follow the above laws and observations.

The largest numbers of migrants came from rural villages close to Brightlingsea. Seventy-three individuals born in Saint Osyth, less than six kilometres away, moved to the town. Likewise, fifty-four from Great Bentley, forty-nine from Thorrington, twenty from Alresford, fourteen from Frating and eleven from Mersea all moved from within a six-kilometre radius. In total 224 individuals are shown as having migrated from within the six-kilometre radius, and a further 216 from within ten kilometres. The latter figure includes migrants not only from rural settlements but also forty-three from the town of Clacton and forty-two from Colchester, the largest town in

Ladysmith Avenue, early 1900s.

the area. A further 131 migrants originated from within ten to twenty kilometres, sixteen between twenty to thirty kilometres, and nine coming from further than thirty kilometres. A further nine individuals originated from areas that are unknown within Essex. Fig.8 summarises this data.

Migrants from beyond Essex, but within Great Britain, total 217 individuals. It is interesting to note that a large group of these (forty-one) came from London. Another factor that becomes strikingly apparent is that considerable numbers of migrants originated from more distant counties and ports known to be used by Brightlingsea's mariners.[145] These include Flushing and Falmouth (Cornwall), Plymouth (Devon), Portsmouth (Hampshire), Bosham and Emsworth, (Hampshire, mentioned earlier), Newhaven and Shoreham (Sussex) and Dover (Kent). Along the east coast this includes the Suffolk ports of Lowestoft, Woodbridge and Ipswich, in addition to Wells and Yarmouth in Norfolk.

This wider migrational pattern which (outside Essex) favours fishing communities as its source appears to confirm that Brightlingsea's developing maritime trade exerted a considerable 'pull' factor for more specialised migrants. It would have been these communities that supplied labour which was already experienced and skilled. They should also have known of Brightlingsea's increasing work opportunities at greater distances as a result of the town's mariners putting into ports on the south and east coast. News and information must have disseminated by word of mouth throughout the trade via the 'mariner's grapevine'. Had the maritime opportunities been less favourable in these more distant ports, then with the knowledge of a good trade at Brightlingsea a migrational 'push' would clearly have been felt by those with the skills to exploit it. Fig.9 summarises migration to Brightlingsea from outside Essex.

Fig.9 is arranged with the counties nearest to Brightlingsea to the left, and those furthest away to the right. This clearly illustrates that the counties of Norfolk, Sussex and Hampshire (with their coastal ports) contributed relatively high numbers of migrants to Brightlingsea despite being further away than counties such as Hertfordshire and Cambridgeshire where fewer migrants originated. During this period Brightlingsea also received migrants from outside the UK mainland into its marine trade, although these numbers are minimal: ten in total. Six of these came from Ireland, two from Jersey, one from Bermuda and one from Australia. Finally, the data reveals that all migrants coming to enter Brightlingsea's marine occupations were male.

To summarise the above, Brightlingsea's population during the mid- to late nineteenth century appears to grow at a rate exceeding that of many similar fishing communities, and is more typical of a larger port of the time. Although Brightlingsea had a roughly even male to female ratio, the data suggest that the marine trades of the town were essentially male occupations, and the migrants to those trades were therefore mostly male. Domestic and craft sectors were more dominated by female

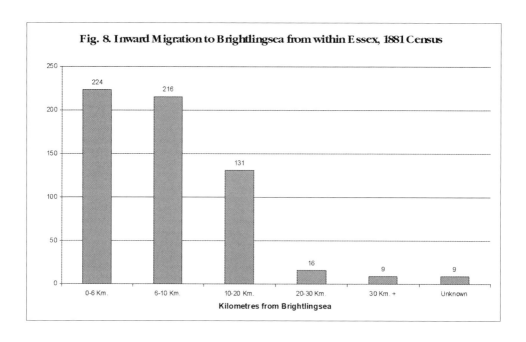

Fig. 8. Inward Migration to Brightlingsea from within Essex, 1881 Census

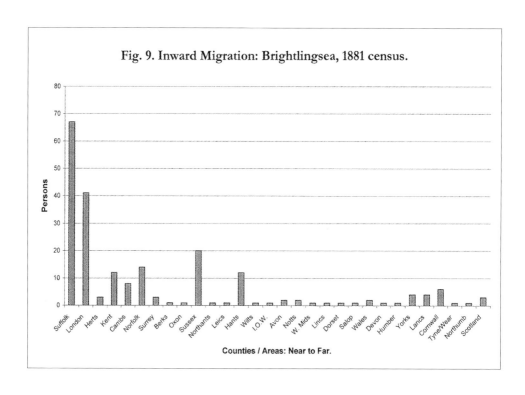

Fig. 9. Inward Migration: Brightlingsea, 1881 census.

Above and below: Victoria Place and the chapel-of-ease, *c.*1900.

Looking from the causeway into the creek, with smacks moored.

labour, with professional and retail areas following a similar trend towards the turn of the century.

The marine trades throughout the period of study are shown to have occupied a relatively high proportion of the town's labour force. Migration to Brightlingsea from within Essex seems to conform to Ravenstein's principles: the majority travelled the shortest distances, and originated from surrounding rural areas; although if we consider the town itself to be a rural community because of the continuing importance of local farming then this situation echoes the findings of G. Nair and D. Poyner described earlier in this chapter. However, migration from outside the county conforms less with this model, with considerable numbers originating from more distant fishing communities. This tends to confirm the 'pull' factor of Brightlingsea's buoyant maritime trade and a 'push' from less prosperous fishing communities as an awareness of Brightlingsea's growing trade spread around the English coastline.

CHAPTER FIVE

Housing and Physical Development

Having examined Brightlingsea's population trends, this chapter turns to the town's nineteenth-century physical development. The growth of domestic housing and the adaptation of existing property will be considered, as well as the development of inns, hostels, boarding houses, retail premises, schools, religious buildings, the premises of voluntary organisations and municipal buildings.

Britain's rapidly growing population in the nineteenth century impacted upon towns of all sizes causing rapid structural growth fuelled by economic development.[146] The physical nature of this impact upon smaller towns would manifest itself primarily in three ways. Firstly, existing buildings could be transformed and adapted to suit changing needs or growing demands. Secondly, buildings could be demolished in order to make way for more up-to-date structures, particularly if the original was in poor condition, (quite often it was more practical to simply knock down and start again, rather than attempt to improve an already dilapidated building). Finally, new buildings could be built on land previously undeveloped.[147] Given the rapid expansion of towns at this time, this third process was a common option in places where nearby land was available. Brightlingsea was not alone in having surrounding land which could be devoted to new development and urban spread: the residents of the London suburb of Camberwell are said to have been amongst many in such places during the nineteenth century who 'could measure the retreat of the open country in terms of furlongs per year'.[148]

Brightlingsea in 1800 could scarcely be described as a town – village would be a more appropriate term. But by the end of the century a town it most certainly was, with all its identifying physical, social and economic characteristics firmly in place. Its development was, therefore, a rapid process that typified the situation in many small towns and suburbs throughout Britain during the mid-nineteenth century where local economic conditions would prove attractive to the now mobile workforce.

Chapman and André's map of 1777 gives a good idea of the extent of housing at Brightlingsea by the end of the eighteenth century.[149] The medieval 'town' is described as a typical Essex polyfocal settlement, with the church forming a focus in the north-west and Hearse Green and North End Green forming further foci to the south. A network of lanes connects these and several outlying farmsteads.[150] The southern cluster of houses, close to the waterfront, must be the area described by Major Thomas Reynolds (see Chapter One) as the 'straggling village, inhabited by oyster dredgers and their families'. The road leading into the triangle of present-day Hurst Green where the majority of these early houses are shown in Chapman and André's map today has some of the oldest properties in the town, confirming this to be the original centre of the community. The plaque on one grade II listed building claims that it was built around 1570. Another, Jacobs Hall, is almost certainly an open-hall medieval structure built over 600 years ago.[151]

Some of the tracks leading from these early houses on the higher ground down to the waterfront are still visible. The road leading from Hearse Green in a westerly direction, known today as the High Street, appears to have been at the heart of the early community: its situation, certainly, would have been advantageous to the early oyster dredgers, being reasonably close to the water but far enough (and high enough) away to avoid the problems associated with high tides and extreme weather.

Jacobs Hall: view from the garden.

Jacobs Hall from the east front.

Doorway, Jacobs Hall.

Left: Jacobs Hall interior.

Below: Jacobs Hall interior, detail of ceiling.

Jacobs Hall interior, the fireplace.

Jacobs Hall Restaurant.

Hurst Green, early 1900s.

With a population recorded at less than 1,000 people in 1801, the area along the present-day High Street would have been little more than a small hamlet surrounded by outlying farms, some distance from All Saints church. As the population grew in the early years of the nineteenth century this central community expanded at the western end of the High Street (then Street Green),[152] now the present-day town centre, and in the east around Hurst Green. Several houses from this period are still standing, contrasting with the later nineteenth- and twentieth-century buildings.

By 1841, with a population of just over 2,000 people (of whom the majority of workers were in marine trades), the community was centred on a residential arc extending from around Hurst Green in the east, along the High Street, to the expanding western end. The building of houses thus far seems to have occurred more towards the north-western end of the existing community, towards All Saints church and the road out of Brightlingsea. The extent of the mid-nineteenth-century community and its development since the late eighteenth century is shown in the 1841 tithe map of the town.[153] Outlying farms and their fields still surrounded a small central settlement. Even the area to the south of the residential area, down to the waterfront, is at this time still largely agricultural.

Brightlingsea's population growth during the second half of the nineteenth century, as detailed in the previous chapter, created a need for housing and commercial premises that began to consume more surrounding agricultural and common land. The proposed new railway line linking the town to Wivenhoe and Colchester was

Above: The High Street looking west, 1905.

Right: Brightlingsea tithe map of 1841.

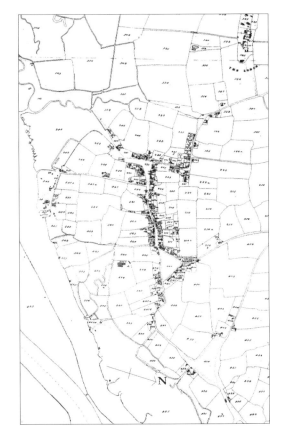

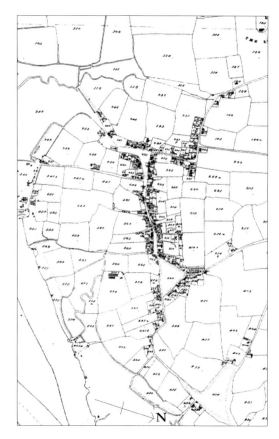

Left: Details from tithe map of 1841.

Opposite: Brightlingsea Station, early 1920s.

probably the catalyst for a large number of plots of building land to be offered up for sale in 1860. Plans from an auction of this date show plots numbering up to 113, with many marked as sold, situated at the south-western end of the High Street, in an area (roughly speaking) formerly known as Gandergoose Green.[154] These were advertised as being close to the proposed new railway station.[155] Documents show that some of this land was owned by James Robinson, based in Pontefract, Yorkshire.[156] Indentures from this date also show that land on this side of the town formed part of the estate of the Dorrien Magens, a well-known banking family living at Hammerwood Lodge in Surrey. John Dorrien Magens inherited a good deal of property here, including Brightlingsea Lodge, Lodge Farm, Hall Farm, Moverons Farm, 'accommodation land', 'Herony and woods' and freehold oyster layings and pits.[157] As his interests lay mainly outside Brightlingsea, much of this land was gradually made available as building plots and to accommodate the new railway line.[158] With undeveloped land such as this available and adjoining the existing town it was, therefore, inevitable, given the economic conditions and needs of the growing community, that a considerable number of properties would be built here.

Indeed, over the next fifteen years, with the railway link through to Colchester and London finally being opened in April 1866, more domestic house building occurred, which led to the formation of New Street and Sidney Street.[159] These two new roads, lined with fairly typical terraced and semi-detached houses, ran from the High Street in a southerly direction to the waterfront.

The illustration overleaf shows details from an advertisement for an auction of building plots at Brightlingsea in 1860.[160] It typifies several such advertisements and catalogues from that time. This particular document conveys much about how the town's potential physical and economic expansion was perceived: 'there appearing to be a great demand in Brightlingsea for eligible Freehold land suitable for building purposes', that the town 'would also afford facilities for sea-side visitors' and that 'Docks and Wharves... are also projected'.

In addition to the White Lion Inn at Hearse Green, the Railway Hotel close to the new railway terminus, the Park Hotel and the Cherry Tree Inn to the north-west, three more public houses (the Freemasons Arms, the Star of Denmark and the Yachtsman's Arms) were now established along this new area of development. The waterfront area underwent further expansion with the building of a gas works, warehouses and two smithies near the existing boatyards and wharf. Many of the buildings detailed above are shown on the 6in Ordnance Survey map of the town from 1875.[161]

Also evident in this illustration are brick-fields and kilns to the north of the railway terminus and south of Hearse Green. The appearance of these was symptomatic of the need to supply locally produced bricks for the new developments. Correspondence

At the Duke of Wellington Inn, Brightlingsea,

On FRIDAY, DECEMBER 14th, 1860,

AT TWO O'CLOCK IN THE AFTERNOON,

IN EIGHTY LOTS.

Particulars of Sale, (with Plans) may be had, and the Conditions inspected, at the Offices of Messrs. DUDDY and SON, 90, Chancery Lane, London, and Maldon, Essex; of G. BRADLEY, Esq., Castleford, Yorkshire; of Mr. G. RUFFELL, Land Surveyor, Brightlingsea; and of the Auctioneer, Trinity Street, Colchester.

PARTICULARS.

The Building Lots offered for Sale in October last, having been all disposed of at the Auction to various purchasers, and there appearing to be a great demand in Brightlingsea for eligible Freehold land suitable for Building purposes, the Vendor has determined to dispose of, by public competition, a further portion of this beautiful and admirably situated property.

The first portion of the Lots now offered immediately adjoin the Lots sold at the previous Sale, and are equally well situated in the Lodge Park; commanding fine views of the German Ocean and mouth of the Harbour. For convenience of access, roadways 30 feet wide are marked out. The other Lots are still more conveniently situated, most of them abutting upon a roadway of 42 feet width, leading directly to the waterside, which roadway materially lessens the distance from this part of the Town to the water and waterside premises.

It is proposed that the terminus of the Railway shall immediately adjoin the Southern part of this road, so that it may hereafter become one of the principal thoroughfares in the Tendring Hundred. Since the last Sale, the requisite notices have been given for an application to Parliament during the ensuing Session, and the necessary steps are being taken to obtain the Act. When the Act is obtained, the line will be proceeded with as quickly as practicable, and its completion may be expected within two years of the present time. This and the Wivenhoe line, when completed, will place Brightlingsea within twenty minutes ride of Colchester, and will also afford facilities for sea-side visitors, and materially increase the value of the property. Docks and Wharves near the Southern end of the Waterside road are also projected, to enable Merchants, Shipowners and others to carry on their business at Brightlingsea, and for convenience of Landing and Carriage of Coal, Fish, Fish manure, Chalk, and other merchandise. The Creek affords an excellent harbourage for Yachts in winter, and is justly celebrated for its Oysters. Being within the ancient jurisdiction of the Cinque Ports, the inhabitants of the Town are entitled to many privileges. An abundant supply of excellent water is obtained from the Spring immediately adjoining this property; from which, nearly the whole of the inhabitants derive their supply.

(*N.B.—The Lots now offered are for convenience, numbered in continuation of the Lots sold at the first Auction; this Sale therefore commences with Lot 33.*)

Lot 33.—A very desirable Building Site, in a commanding situation in the Lodge Park having a double frontage, viz.—60 feet to the Park Road, and of 120 feet to the Park Chase as shown on plan.

Lot 34.—The adjoining Site of similar dimensions, frontage 60 feet, depth 120 feet.

Lot 35.—A similar Site, frontage 60 feet, depth 120 feet.

Lot 36.—A similar Site, ,, ,, ,,

Lot 37.—A similar Site, ,, ,, ,,

Lot 38.—A similar Site, ,, ,, ,,

Lot 39.—A similar Site, ,, ,, ,,

Lot 40.—A similar Site, ,, ,, ,,

Left: Building plots advertised for sale, 1860.

Below: Brightlingsea from the 1875 Ordnance Survey.

between the Aldous Company (boat builders and marine engineers) and the Brightlingsea Urban District Council shows that industrial development at the wharf continued up until 1912.[162]

Domestic housing continued to grow for the remainder of the nineteenth century and well into the twentieth. Expansion of the mainly terraced and semi-detached housing to the south led to the creation of Tower Street and Lime Street, running parallel and east of Sidney Street. Lime Street, following an earlier pathway to the waterfront, remains un-metalled to this day. A grand new public house, The Anchor, appeared at the waterfront close to the now well-established shipbuilding yards. To the west, around the housing plots developed after 1861, further development occurred from the railway terminus northward through the 'Park' area, which included another new public house, The Duke of Wellington. Building development during the twentieth century was mainly centred in the north-west of the town.[163] The map opposite illustrates the development of Brightlingsea from 1800.

During the decade 1861–71, domestic houses were being constructed in Great Britain at an average rate of approximately 65,000 each year. By the decade 1891–1901 the figure was approximately 126,000, representing an increase of 94 per cent.[164] This increase, created by the needs of the growing population, was mainly centred on developing industrial towns and cities as opposed to rural areas. The evidence so far

From the Aldous Heritage Dock, looking toward new flats.

Left and below: From the Aldous Heritage Dock, facing toward Point Clear.

Right: Smacks sheeted over at the Aldous Heritage Dock.

Below: The estuary, 2009.

The old Anchor building and the Colne Yacht Club today.

Above and below: The Anchor Hotel.

suggests that Brightlingsea's total housing stock more than doubled in the second half of the nineteenth century, and continued to grow in the early twentieth. An important factor also apparent from this is that a large proportion of houses built during the mid- to late nineteenth century were workers' dwellings, built at the southern end of town close to the waterfront at the hub of the maritime trade. The properties built to the north-west at this time were more substantial and of higher status. Census data from 1881 and 1901 shows that a very high proportion of the workers' dwellings in the south, including New Street, Sidney Street and Tower Street, close to the waterfront, were occupied by maritime workers and their families, such as George and Tamara Wringe and their children, and Joseph and Maria Taylor.[165]

Brightlingsea's rapid physical growth and development patterns therefore appear to be more typical of larger towns, suburbs and even industrial centres throughout Britain at this time, and the town as it stands today reflects this clearly. The following observation of such towns in England, made during the 1960s, has a remarkable affinity with modern Brightlingsea:

> The High Street, where the main shops are, will probably be a longish street. It will almost certainly be irregular in its configuration. Here and there, there will be a jut or recession in the building lines. Somewhere along its course there will be a swelling or a narrowing of the roadway. The two sides of the street will not be exactly parallel; and in their alignments, though not in the kinds of buildings that front them, they may even

New Street, early 1900s.

Right: New Street, 2009.

Below: Tower Street, early 1900s.

seem to have a curious independence of each other. Sometimes there may be a break among the buildings for the forecourt of the council offices, or a chapel or something of that kind: and a tree may lean out there with a happy effect on the scene.[166]

Several of the original buildings in the oldest part of the town, along the High Street and Hurst Green, are still standing, but many have been demolished and replaced by nineteenth- and twentieth-century properties.

The High Street also seems to have been the centre for shops and non-marine businesses: advertisements taken from the *BPM* of 1884 give examples of these businesses, while other advertisements show shops and businesses in roads leading directly from the High Street.

Silcott Street, early 1900s.

Silcott Street, 2009.

The post office, Victoria Place.

The former post office building.

Victoria Place.

The Promenade and Bateman's Tower.

Victoria Place and the King's Head Hotel.

Houses in New
Street, 2009.

Left: High Street, 2009.

Below: The Kings Head, 2009.

This central area remained the commercial focus well into the twentieth century.[167] The High Street shops are typified by much re-development and alteration during and after the late nineteenth century. Some evidence of this is apparent in a visual survey. Other commercial premises based on the marine trade, such as the shipbuilding yards, smithies, oyster sheds and chandleries developed and expanded along the waterfront. The new gas works, mentioned earlier, also appeared at the industrial end of town.[168]

Of course Brightlingsea's growth was not limited to just commercial and residential building: religious and public buildings also increased in number, as one might expect in a vibrant and developing community. The period 1830–90 saw the construction of an Anglican chapel-of-ease in the centre of the town, while elsewhere Wesleyan and Primitive Methodist chapels, a Congregational chapel and an impressive New Church chapel appeared. In addition to these the town gained a Salvation Army headquarters. Secular and educational buildings were also constructed during this period, including a National school, a YMCA hall, an Ancient Order of Foresters hall and a Masonic temple, the social aspects of this having been considered earlier.

The above evidence shows that Brightlingsea's pattern of physical growth during the nineteenth century had much in common with that of urban and suburban areas throughout the country that either had or were close to expanding industries. This includes expansion into undeveloped surrounding land, the provision of workers' housing, a connection to the rail network, an expanding industrial area, growing numbers of retail premises in the town centre and increasing numbers of buildings to serve the religious, educational and social needs of the growing population. This economic growth was anticipated by landowners who made plots of building land available as the town developed. A healthy local economy and the extension of the branch line from Wivenhoe would make this land even more valuable as the demand for housing grew. The overall pattern of growth resulted in an industrial centre based along the waterfront with a retail, civic and social focus along the High Street.

Advertisements in *BPM* for High Street businesses, 1884.

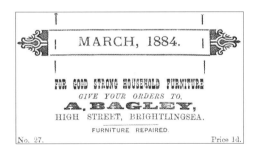

MARCH, 1884.

FOR GOOD STRONG HOUSEHOLD FURNITURE
GIVE YOUR ORDERS TO,
A. BAGLEY,
HIGH STREET, BRIGHTLINGSEA.

FURNITURE REPAIRED.

No. 27.   Price 1d.

W. BISHOP,
BOOT and SHOE WAREHOUSE,
High Street, Brightlingsea,

AGENT for the Indestructible Toes for Boys' and Girls' Boots, guaranteed not to wear through during the existence of the Boots.
N.B.—Selling Off at Reduced Prices for Cash ! !

W. WARREN,

BEGS to inform the inhabitants of Brightlingsea that he has Opened a Branch Shop in the High Street, for the supply of Good Strong Home-Made Boots and Shoes, Unequalled in style. Excellent in quality. At Manufacturers' Prices, for Cash. Orders taken, and repairing done.
☞ *Please note the Address—*
HIGH St., BRIGHTLINGSEA.

The Blacksmith's,Victoria Place 1895.

Towards the waterfront.

The High Street and St James's, 2009.

The High Street and Victoria Place, 2009.

Nelson Street 2009.

Spring Road, early 1900s.

The 'Old Cottage'.

East End Green, 1905.

CHAPTER SIX

Conclusions

During the nineteenth century Brightlingsea entered a period of considerable physical, economic and social development, coupled with a rapid growth in population. Comparisons with other ports show that Brightlingsea's experience was not typical: some ports of similar size did not witness such growth, and indeed this level of expansion was more typical of a larger contemporary port or industrial suburb. The proportion of residents engaged in maritime trades and the pattern of the town's physical development suggest that this 'economic boom' was driven by the town's expanding marine industry. This industry, as has been shown, was an essentially male occupation and centred primarily, but not exclusively, on the oyster trade.

Other existing or new industries and occupations, with the exception of agriculture, were effectively service industries that developed as a result of the town's economic success; they in turn may have created growth, but did not instigate it. This economic success created a migrationary 'pull' not just from the immediate locality but from farther afield as well. On a wider scale, inward migration was higher from more distant fishing centres than from nearer non-maritime areas, a fact which partly challenges Ravenstein's principles. More distant ports where migration originated did not necessarily experience such levels of growth, generally having more static populations, which suggest that the industry at Brightlingsea was uncommon and specialised. The overwhelming numbers of male migrants also challenges Ravenstein's principles.

Much of the physical development at Brightlingsea from the mid- to late nineteenth century consisted of workers' housing, in particular the creation of New Street, Sydney Street and Tower Street. The waterfront also saw considerable expansion with the development of shipyards and industrial workshops. The arrival of the railway was also a major factor in encouraging further development. Census data shows that residents

Right: The Old Wreck Warehouse, 2009.

Below: The old Anchor building, 2009.

The new marina area, 2009; formerly the James and Stone Shipyard.

The smack dock.

in the new housing were largely mariners or engaged in marine trades, which confirms that these trades were central to the town's economic growth.

Much of this development was made possible by the availability of land. As demand grew land was made available by landlords who were clearly happy to capitalise on the situation. Indeed, most of the land was of manorial origin and in the possession of absentee landlords.

The trade in oysters that appears to have been central to Brightlingsea's development was clearly affected by their popularity across the social spectrum. Demand for oysters rose nationally, in particular from London. Being close to the capital by sea and rail ensured that Brightlingsea became one of the key ports for this trade. Other vital factors were the expertise of the town's shipwrights in building sturdy smacks and the experience and skill of the oyster dredgers working offshore. Other maritime activities were also important: indeed, there is a strong school of thought that emphasises pleasure boating and yacht racing as major economic factors, but primary evidence clearly suggests that if this is the case it is only in addition to the thriving oyster trade.

When demand for oysters waned, largely as a result of typhoid scares, but also because of over-dredging and the hazards of working offshore in winter, the Brightlingsea mariners were able to turn to other activities such as scallop dredging and stow-boating for sprats. Also, as mentioned, towards the end of the nineteenth century Brightlingsea, having gained a good reputation for shipbuilding and seamanship, saw

The Causeway and The Hard, early twentieth century.

The old Anchor building and the Colne Yacht Club today.

an increasing number of ocean-going racing and pleasure yachts being berthed in the harbour, a development which provided additional seasonal work for the mariners. This is confirmed by the appearance in the 1901 census of the occupations 'yacht steward' and 'yacht captain', as well as by the news given in *BPM*.

The growth of Brightlingsea's population and economy clearly created an environment in which cultural and religious aspiration could flourish. A dramatic rise of nonconformity within the community occurred as new 'lower middle-class' marine workers and their families arrived, in a pattern similar to that of East-Anglian 'craft villages'. Congregations were meeting close to the centre of town, so, not to be outdone, the Anglican community built a central and substantial chapel-of-ease on the High Street.

Opportunities, driven by growing prosperity, also increased. The new National and Sunday schools were highly successful, gaining national recognition. As the nineteenth century progressed poverty appears to have decreased within the community, and the thriving oyster trade also created a need for an increased police presence. Philanthropic organisations and societies proliferated and the town began to acquire all the characteristics of a thriving Victorian town or suburb.

Towards the end of the nineteenth century Brightlingsea saw the arrival of key individuals who embraced the community and promoted its 'urban status'. Having become an urban district, the town was further enhanced by the revival of the Cinque Port connection. The manorial courts, although less important in the business

The Hard in 1885 with the coastguard hulk.

Lower Green, 1890 (now Victoria Place).

Stormy weather, Bateman's Tower.

they conducted as compared to the previous century, were nevertheless maintained almost as a matter of civic pride. It has been suggested that the town aspired to become a borough with its own corporation; but as the economy of the town peaked around the turn of the century any possibility of this receded. Although the marine trades adapted to survive, and undoubtedly benefited from the arrival of the racing and pleasure yachts, the level of growth and development seen in previous decades could not be sustained.

This investigation has raised questions for me (and I hope will provoke further discussion and enquiry elsewhere), and exposed areas in need of further research: for example, the nature of the relationship between the mariners and non-conformity (particularly the New Jerusalem church) not just at the local level but on a national scale. At first sight one might think that Brightlingsea's development in the nineteenth century fits into the normal pattern of sustained urban growth demographically, economically and socially. It began the century as an agricultural parish of scattered settlements which contained a small fishing hamlet at its seaward end. By 1900 that hamlet had grown into a small but significant town-cum-port, based largely on a boom in the offshore oyster trade. However, the boom was not sustained much beyond the turn of the century and the town's aspirations to become an incorporated borough were never realised. This theme of short-term, limited urbanism in the nineteenth and twentieth centuries clearly applies to many other English communities, and undoubtedly deserves greater attention from historians of the future.

GLOSSARY

Beam   The widest point of a ship or boat.

Brood Oysters   Young oysters that can be transplanted to continue growth elsewhere.

Bumkin   An inshore fishing boat under sail, smaller than a smack.

Chandler   A purveyor of boat fittings, equipment and consumables.

Cinque Ports   The five south-coast medieval ports charged with building ships for the Crown.

Common Grounds   Areas of inshore seabed that can be dredged for oysters by anyone.

Costermonger   A London street trader who buys food or general goods wholesale and sells them on to the public.

Cutter/Cutter Rig   A vessel carrying a main mast rigged with two headsails in addition to mainsails.

Dredge   An apparatus designed to be dragged along the seabed to gather oysters. To dredge for oysters is to carry out this activity.

Frames   Timber formers or 'ribs' onto which the hull planks are fixed.

Forepeak   The area below deck at the front of a boat.

Gaff Rig   A triangular topsail supported by a spar at the top of the mainmast.

Garboard Strake   The length of plank above the keel.

Keel   The 'backbone' of a vessel from which rise the frames and planking.

Layings   Areas of inshore seabed used to grow oysters, until ready for harvest.

Oyster Beds   Small lagoons, sometimes naturally occurring but normally cut into the foreshore for growing young oysters until ready for transplanting to layings.

Shipwright   A skilled worker who builds and repairs ships and boats.

Skiff   A small, light rowing boat.

Skilling / Skillingers   A corruption of 'Terschelling', a word used to describe crew and vessels who dredged for oysters in this notorious area near the Dutch coast, and the activity itself.

Smack   A fishing vessel under sail engaged in taking fresh fish.

Spat   Oyster larvae that have settled and started to grow into young oysters.

Stow-boating   'Stalling' a vessel into the tide, with nets slung below to catch fish.

Tide Waiter   An individual who awaits vessels to charge duty on cargo.

Wet Well   A tank within a vessel that allows fresh sea water to flow through.

Wharf   A harbour area where vessels may come alongside to load or unload.

BIBLIOGRAPHY

Primary Sources

ERO denotes a source from the Essex Records Office, Chelmsford. **ERO/COL** denotes a source accessed at the Colchester office prior to its closure at the end of March 2007.

For printed sources, the place of publication is London, unless otherwise stated.

The National Archives:

Oyster company files (1882), BT31/3022/17094.
Harbour reports (1865–1917), MT10/186.

ERO:

Census enumerator's books, Brightlingsea, microfilm (1841, 1851, 1881 and 1901).
The Essex Standard (1841–42), T/A 572/7.
Miscellaneous documents, Colchester / Brightlingsea (1420–1812), T/P 1/14.
Sale of shipyards, buildings and land (1886), ACC.C.32.
Admission register (1883–97), D/Q 40/2.
Plan, Brightlingsea pier (1867), D/DU 210/36.
Certificates of ordination (1861–72), D/P 312/2/1.
Court book (1826–39), D/DU 210/10.
Court book (1864–80), D/DU 210/14.
Select vestry orders for relief of poor (1827–28), D/P 312/8/1.
Indentures (1653–1833), D/P 312/14/1.

Minute books (1842–75), D/P 312/8/4.
Sequestration order (1878), D/P 312/7/1.
Deed of gift of chain of office (1892), D/DCm/Z1A.
Plan for Salvation Army hall (1908), D/UBR PB21/15.
Rural Dean's parochial return (1844), D/P 138/28/2.

ERO/COL:

Extracts, transcripts: Brightlingsea (1700–1909), T/P 1/15.
Brightlingsea tithe map (1841), D/CT 51/B.
Sale catalogues (1795–1886), D/DEL E26.
Building plots for sale, indentures (1828–60), D/DCfT 51.
Building plans (1912–15), D/UBR PB23/3.
Building plans (1907), D/UBR PB21/4.
6in O.S. maps (1875, 1897, 1923).
Consecration of chapel-of-ease (1836), D/P 8/1/16.
Prisoners accounts, county gaol (1828–34), D/P 312/18/7.
Chapel book (1822), D/NM 1/2.
Accounts (1822–57), D/NM 1/1.
Plans for Wesleyan school (1861), E/P 15/1.
Quarter sessions, registered place of worship (1840–41), Q/SBb 542/20.
Church book (1846–94), D/NC 30/1.
Church circular (1889), D/NC 30/19.
Congregational church register (1891–1904), D/NC 39/16

Primary Sources, Printed:

25in O.S. map of Brightlingsea (1874), courtesy of Mr Guy Smith.
Mitchell, B. *Abstract of British Historical Statistics* (Cambridge, 1962).
CUL/Atlas.2.97.6. Chapman and André's map (1777), plates xiv and xv.
Baker, J. *Introduction of Wesleyan Methodism into Brightlingsea* (1884), ERO Transcript No. 295.
Ancient Order of Foresters Directory (1967).
Essex Standard (30 March 1864).
Brightlingsea Parish Magazine (1882–1914)
Hore, J. and Jex, E. *The Deterioration of Oyster and Trawl Fisheries of England: Its Cause and Remedy* (1880).
Essex Record Office Publications No. 30, *Essex and The Sea* (Chelmsford, 1970).
1851 Census of Great Britain, *Population Tables* 2, Table [1], 'Population Abstract'.
Wilson, J. *Imperial Gazetteer of England and Wales,* 1,2 (Edinburgh, 1874).
Youngs, F. *Local Administrative Units in England and Wales,* 1 (1979).
Brightlingsea Conservation Area – Final Report, *Tendring District Council Conservation Area Review* (2006).

Crockford, Clerical Directory

Mayhew, H. *London Labour and the London Poor* (1851).

White, W. *Directory of Essex* (Sheffield, 1848).

House of Commons Accounts and Papers, V. 16 (1854), Vol. LIV; LXIV (Shipping) session (1867).

London and Provincial Medical Directory (1854–92).

Internet Sources (Last accessed August 2007):

http://www.brightlingsea-town.co.uk

http://www.visionofbritain.org.uk

http://www.hammerwood-mistral.co.uk

http://www.pioneersailingtrust.org.uk

http://www.cinqueportliberty.co.uk

http://www.csiss.org/classics/content/90

Parliamentary papers on-line (accessed at CUL): 1808 (296) Bill for more effectual protection of oyster fisheries; 1823 (461) Bill for protecting and regulating public oyster fisheries in England and Wales; 1876 (65) Bill for better protection of oyster fisheries; 1862 Committee of Council on Education: report, appendix (1861–1862), 510; 1870, Department of Science and Art: 17th Report, Appendix, 168; 1875, Committee of Council on Education: Report, Appendix, (1874–5) bye-laws of school boards; 1882, Committee of Council on Education: Report, Appendix, 404; 1819 (224) Select Committee on Education of the Poor (1818), Digest, Parochial Returns, 253.

Secondary Sources, Books:

Ashworth, W., *The Genesis of Modern British Town Planning* (1954).

Benham, H., *Essex Gold* (Chelmsford, 1993).

Benham, H., *The Last Stronghold of Sail* (1986).

Dickens, C., *The Posthumous Papers of the Pickwick Club,* 1 (1837).

Dickin, E., *A History of Brightlingsea* (Brightlingsea, 1939).

Dove, C., *The Liberty of Brightlingsea* (Brightlingsea, 1974).

Drummond, J., *The Englishman's Food* (1939).

Dyos, H., (ed.), *The Study of Urban History* (1998).

Dyos, H., *Victorian Suburb* (Leicester, 1961).

Elrington, C., (ed.), *The Victoria County History of Essex,* 9 (1994).

Gordon, D., *A Regional History of the Railways of the Eastern Counties,* 3rd Ed., V. 5, *Eastern Counties* (Newton Abbot, 1990).

Hall, H., *Lucullus: or Palatable Essays, 1* (1878).

Hey, D., (ed.) *The Oxford Companion to Local and Family History* (Oxford, 1996).

Higgs, E., *A Clearer Sense of the Census* (1996).

Higgs, E., *Making Sense of the Census: The Manuscript Returns for England and Wales, 1801–1901* (1989).

Hunt, E., *Regional Wage Variations in Britain: 1850–1914* (Oxford, 1973).

Leather, J., *The Northseamen* (Lavenham, 1971).

Lummis, T., *Occupation and Society in East Anglia, 1880–1914* (Cambridge, 1985).

Marks, R., (*et al.*) *Pioneer, Last of the Skillingers* (Wivenhoe, 2002).

Neild, R., *The English, The French and The Oyster* (1995).

Pooley, C. & Turnbull, J., *Migration and Mobility Since the 18th Century* (1998).

Quennel. P., (ed.), *Mayhew's London* (1949).

Sharp, T., *Town and Townscape* (1968).

Starkey, D., (*et al.*) *England's Sea Fisheries* (2000).

Steel, D., *National Index of Parish Registers Vol. 2, Sources for Nonconformist Genealogy and Family History* (Chichester, 1973).

Wakeling, A., *Brightlingsea Society of the New Church: A History 1813–1968* (1968).

Woods, R., *The Demography of Victorian England and Wales* (Cambridge, 2000).

Wrigley, E., (ed.), *Nineteenth Century Society: Essays in the Use of Quantitative Methods for the Study of Social Data* (Cambridge, 1972).

Wrigley, E. & Schofield, R., *The Population History of England 1541 – 1871: A Reconstruction*, Second Ed. (Cambridge, 1989).

Yonge, C., *Oysters* (1960).

Secondary Sources, Articles:

Corbett, J., 'Ernest George Ravenstein: The Laws of Migration, 1885', in www.csiss.org/classics/content/90, accessed 03:07:06.

Cross, A., 'Old English Local Courts and the Movement for their Reform,' *Michigan Law Review*, 30, No. 3 (1932), 371.

Everitt, A., 'The Pattern of Rural Dissent: the Nineteenth Century', *University of Leicester Dept. of English Local History Occasional Papers*, Second Series, No. 4 (1972).

Lockwood, M., 'Marine Policing in Essex,' *Essex Police History Notebook*, Pt. 1, No. 31.

Long, J., 'Rural – Urban Migration and Socioeconomic Mobility in Victorian Britain', *The Journal of Economic History*, Vol. 65, No. 1 (March 2005), 1–35.

Nair, G. and Poyner, D., 'The Flight from the Land? Rural Migration in South–East Shropshire in the Late Nineteenth Century', *Rural History*, Vol. 17, No. 2 (2006), 167–186.

Ravenstein, E., 'The Laws of Migration', *Journal of the Statistical Society of London*, 48, 2 (June 1885), 167–235.

APPENDIX ONE

The following passage is taken from Henry Mayhew's *London Labour and the London Poor* and gives an unparalleled account of how, and in what quantities, oysters were sold and consumed in the capital during the nineteenth century:

The trade in oysters is unquestionably one of the oldest with which the London – or rather the English – markets are connected; for oysters from Britain were a luxury in ancient Rome.

Oysters are now sold out of the smacks at Billingsgate, and a few at Hungerford. The more expensive kind, such as the real Milton, are never bought by the costermongers, but they buy oysters of a 'good middling quality'. At the commencement of the season these oysters are 14s a 'bushel', but the measure contains from a bushel and a half to two bushels, as it is more or less heaped up. The general price is, however, 9s or 10s, but they *have* been 16s and 18s. The 'big trade' was unknown until 1848, when the very large shelly oysters, the fish inside being very small, were introduced from the Sussex coast. They were sold in Thames Street and by the Borough-market. Their sale was at first enormous. The costermongers distinguished them by the name of 'scuttle-mouths'. One coster informant told me that on Saturdays he not infrequently, with the help of a boy and a girl, cleared 10s by selling these oysters in the streets, disposing of four bags. He thus sold, reckoning twenty-one dozen to the bag, 2,016 oysters [sic.]; and as the price was two for a penny, he took just 4l and 4s by the sale of oysters in the streets in one night. With the scuttle-mouths the costermonger takes no trouble: he throws them into a yard, and dashes a few pails of water over them, and then places them on his barrow, or conveys them to his stall. Some of the better class of costermongers, however, lay down their oysters carefully, giving them oatmeal 'to fatten on'.

In April last, some of the street-sellers of this article established, for the first time, 'oyster-rounds'. These were carried on by costermongers whose business was over at twelve in the day, or a little later; they bought a bushel of scuttle-mouths (never the others), and, in the afternoon, went around with them to poor neighbourhoods, until about six, when they took a stand in some frequented street. Going these oyster-rounds is hard work, I am told, and a boy is generally taken to assist. Monday afternoon is the best time for this trade, when 10*s* is sometimes taken, and 4*s* or 5*s* profit is made. On other evenings only from 1*s* to 5*s* is taken - very rarely the larger sum - as the later the day in the week the smaller is the receipt, owing to the wages of the working classes getting gradually exhausted.

The women who sell oysters in the street, and whose dealings are limited, buy either of the costermongers or at the coal-sheds. But nearly all the men buy at Billingsgate, where as small a quantity as a peck can be had.

An old woman, who had 'seen better days', but had been reduced to keep an oyster stall, gave me the following account of her customers. She had showed much shrewdness in her conversation, but having known better days, she declined to enter upon any conversation concerning her former life:-

'As to my customers, sir,' she said, 'why, indeed, they're all sorts. It's not a very few times that gentlemen (I can call them so because they're mostly so civil) will stop – just as its getting darkish, perhaps – and look at them, and then come up to me and say very quick: "Two penn'orth for a whet." Ah! Some of 'em will look, maybe, like poor parsons down upon their luck, and swallow their oysters as if they was taking poison in a hurry. They'll not touch the bread or butter once in twenty times, but they'll be free with the pepper and vinegar, or, mayhap, they'll say quick and short, "A crust off that." I many a time think *that* two penn'orth is a poor gentleman's dinner. It's the same often – but only half as often, or not half – with a poor lady, with a veil that once was black, over a bonnet to match, and shivering through her shawl. She'll have the same. About two penn'orth is the mark still; it's mostly two penn'orth. My son says, it's because that's the price of a glass of gin, and some persons buy oysters instead – but that's only his joke, sir. It's not the vulgar poor that're our chief customers. There's many of them won't touch oysters, and I've heard some of them say: "The sight on 'em makes me sick; it's like eating snails." The poor girls that walk the streets often buy; some are brazen and vulgar, and often the finest dressed are the vulgarest; at least, I think so; and of those that come to oyster stalls, I'm sure it's the case. Some are shy to such as me, who may, perhaps, call their own mothers to their minds, though it aint many of them that is so. One of them always says that she must keep at least a penny for gin after her oysters. One young woman ran away from my stall once after swallowing one oyster out of six that she'd paid for. I don't know why. Ah! There's many things a person like me sees that one may say, "I don't know why" to; that there is. My heartiest customers, that I serve with the most pleasure, are working people, on a Saturday night. One couple – I think the wife always goes to meet her husband on a Saturday night – has two, or three, or four penn'orth, as happens, and it's pleasant to hear them say, 'Why don't you have another, John?" or, "Do have one or two more, Mary Anne." I've served them that way two or three years. They've no children, I'm pretty sure, for if I say, "Take a few home to the little ones," the wife tosses her head, and says, half vexed and half laughing, "*Such* nonsense." I send out a good many oysters, opened, for people's suppers, and sometimes for supper parties – at least, I suppose so, for there's five or six dozen often ordered. The maid-servants come for them then, and I give them two or three for themselves, and say, jokingly-like, "It's no use offering you any, perhaps, because you'll have plenty that's left." They've mostly one answer: "Don't we wish we may get 'em?" The *very* poor never buy of me, as I told you. A penny buys a loaf, you see, or a ha'porth of bread and a ha'porth of cheese, or a half-pint of beer, with a farthing out. My customers are mostly working people and trades people. Ah! Sir, I wish the parson of the parish, or any parson, sat with me a fortnight; he'd see what life is then. "It's different," a learned man used to say to me – that's long ago – "from what's noticed from the pew or the pulpit." I've missed the gentleman as used to say that, now many years- I don't know how many. I never knew his name. He was drunk now and then, and used to tell me he was an author. I felt for him. A dozen oysters wasn't much for him. We see a deal of the world, sir – yes, a deal. Some, mostly working people, take quantities of pepper with their oysters in cold weather, and say it's to warm them, and no doubt it does; but frosty weather is very bad oyster weather. The oysters gape and die, and then they are not so much as manure. They are very fine this year. I clear 1*s* a day, I think, during the season – at least 1*s*, taking the fine with the wet days, and the week days with the Sundays, though I'm not out then; but, you see, I'm know about here."

The number of oysters sold by costermongers amounts to 124,000,000 a year. These, at four a penny, would realise the large sum of 129,650*l* [sic.] We may therefore safely assume that 125,000*l* is spent yearly in oysters in the streets of London.

APPENDIX TWO

The extraordinary story of the American millionaire-philanthropist Bayard Brown and his association with Brightlingsea during the late nineteenth and early twentieth centuries is a remarkable adjunct to the history of the town's development at this important time, worthy of study in its own right. Much of the detail has been covered by authors dealing with Essex's coastal maritime history, but the article reproduced below, taken from *Wide World Magazine*, and dated May 1907, by F.R. Temple, gives us an insight into the effect that Mr Brown's presence had upon the local community:

This article deals with an extraordinary American Millionaire, who has lived for nearly twenty years on board a palatial steam-yacht moored in an Essex river. Occupying some part of his time by giving away large sums of money to certain proportion of the scores of applicants who besiege him day and night. Sometimes this eccentric gentleman even bestows hundreds of pounds upon lucky persons without the asking! This seemingly incredible state of affairs is described below.

How many readers of the *Wide World Magazine*, possessing an ocean-going steam yacht and blessed with an abundance of riches, would elect to constantly reside on the vessel, moored in one spot, for nearly twenty years – and that spot an anchorage in an exposed East Coast estuary, where the ordinary contrast between the English winter and the attractions of sunnier climes is often strongly accented?

Such is the strange procedure of an American owner, whose steamship, the *Valfreyia*, has long been a familiar object in the River Colne, Essex, where he furnishes what are probably unique contributions both to the annals of yachting, and the records of philanthropy. It is now about nineteen years since Mr Bayard Brown's yacht, the *Lady Torfrida* (afterwards sold to the Grand Duke Michael of Russia, and replaced by the *Valfreyia*), terminated an extended cruise by dropping anchor a few hundred yards of the small Essex seaport of Brightlingsea, where the Colne is about half a mile wide, near its confluence with the North Sea. There Mr Brown has ever since remained, the big yacht being in continual use as a marine house-boat summer and winter.

Valfreyia, originally owned by Sir Wm. Pearce, Bart., Is a fine screw steamer of seven hundred and thirty five tonnes, handsomely appointed, with modern equipment, electric light, auxiliary steam and electric appliances, steam launch, etc., and was constructed at a cost of about forty thousand pounds, apart from the expense of decorating and furnishing. Besides spacious quarters for owner and friends, she provides accommodation for a crew of about three dozen men. Although in the nineteen years since the *Valfreyia* was launched there have been great developments in yacht construction, as in all branches of marine architecture, there are today but few privately owned vessels of larger size afloat; and among the fleet of big yachts seen every season on the Colne yachting station none presents more graceful or smarter lines.

For a long time after Mr Brown's arrival steam was constantly kept up and a full crew maintained, ready to start for any destination at the shortest notice. These unusual circumstances, with others, directed public attention to the yacht, and numerous stories were in circulation, many of them unfounded rumours, with regard to the vessel and its owner. But when the scale on which Mr Brown remunerated persons who rendered him little services ashore, and distributed largess among even those who did not, became know, passing attention quickly developed into eager interest and speedily culminated in popular excitement.

People of the district almost tumbled over one another in their rush for the millionaire's yacht, and not a few received substantial donations, in some cases amounts largely exceeding their highest expectations. An immediate result was the advent of hold hunters not only from very many districts of Essex and Suffolk, but from London and places even more distant, all intent on obtaining relief from financial embarrassment or the immediate exchange of a daily routine more or less uncongenial for a future of leisured affluence.

These included visitors who tenaciously claimed kin with the millionaire, some of them taking up temporary residence in the locality to press the claim, and proving a source of annoyance through their unfortunate delusions.

It is not surprising to find, therefore, that the visits of the suppliants were after a little time severely, if somewhat capriciously, discouraged. Local boatmen received subsidies of one pound a week each, on condition that they did not ferry people to the yacht, and for days, even weeks at a time, the *Valfreyia's* owner would take no notice of the flotilla of boats in waiting. This soon weeded out the long distance applicants, who, with all their zeal, could not afford the uncertainty of the game, even though aware that, occasionally, Mr Brown, relenting, gave distributions of cash on a more than lavish scale.

But the pursuit of the millionaire by many people of the adjacent districts, and by numbers from a greater distance who have managed to get a place on the list of those to

View from the Colne Yacht Club, 2009.

The new property development from the Colne Yacht Club.

whose personal applications he responds (occasionally, if not regularly), continues till the present time. And so, while the conveyance of suppliants to the yachts forms the chief occupation of most of the local boatmen, the pilgrimage to the *Valfreyia* to solicit the bounty of its owner has long been one of regular industries of the district.

A very large number of persons have participated for years in Mr Brown's benefactions, and to a surprising extent. Nor are they simply the poor and destitute. They include labourers, boatmen, gardeners, artificers, and foremen in various trades, yachtsmen and sailors, ex-public servants (and wives and other female relatives of these), with a liberal admixture of publicans and innkeepers, shopkeepers and other traders, and a sprinkling of farmers, salaried officers of large corporations, etc. Even professional men and persons who are owners of freehold property of material extent have received cash gifts of large amount.

Many cottagers have long been in receipt of one pound and more per week, while some of the other classes get several pounds – even ten pounds and upwards frequently – and all occasionally have extra donations of considerably greater sums. Local agricultural labourers earning thirteen shillings a week have been handed as much as fifty pounds at one time – equal to eighteen months wages; donations to other astonished recipients have run into hundreds at a time. Some of these sums, incredible as it may sound, have been given unsolicited to traders and professional men with whom the millionaire has had some personal dealings. To a local curate he sent three hundred pounds, intended as a wedding present; and there are various substantial farmers and others in the immediate neighbourhood who admit receiving from him in two or three recent years several hundred pounds or more each. Some have even got as much as a thousand pounds at one time.

Most of Mr Brown's disbursements are on a scale of princely munificence. The captain of his yacht, apart from periodical gifts, has for many years enjoyed a salary of one thousand pounds per annum, and it may, perhaps not be without interest to mention that the earliest of the transactions which transformed the *Valfreyia* into a floating gold mine in local estimation was the payment by the owner of ten pounds to a small innkeeper for driving him some three miles to the shore at the close of an evening walk.

At one time the sums donated on board the yacht must have reached two thousand pounds monthly. They are now much less, but it has been computed that, since his arrival in the Colne, Mr Brown's distributions to individuals have aggregated something like a quarter of million pounds sterling.

Practically every day, summer and winter, some boats are ranged round the yacht; sometimes twenty of thirty are assembled at one time, containing sixty or seventy people. Of these the majority are women some of whom walk five or six miles to the shore, hire a boatman (paid partly by results), and trudge home again afterwards; and this they will do on several consecutive days, even in bad weather, until, in their own phraseology, they 'get paid'.

Early in the afternoon the boats assemble in groups round the yacht the occupants prepared with wraps and refreshments for the contingency of a protracted wait. From the shore the boatmen watch for signs on the yacht – distant some three or four hundred yards – that the owner is astir, while their prospective passengers shelter close at hand.

Mr Brown is not an upholder of the early rising convention, and, like his celebrated countryman, Mr T. A. Edison, does not see why people go to bed merely because it is night; so the afternoon is sometimes very far advanced before he quits his private apartments. If he has not appeared by five or six o'clock the curious assembly begins to disperse, some having trains and other conveyances to catch, but a number will occasionally remain for hours later.

When the millionaire does appear on deck in the afternoon he may cause it to be announced that he has nothing to give, and act accordingly for a time, at least; or he may remain apparently oblivious of the presence of the flotilla. Perhaps he will at once approach the side and raise his hat to the suitors, when a number of the more favoured applicants then ascent the gangway in Indian file, receive their gifts from him personally, make their bows, and descent. This is the work of a few minutes. The gangway is drawn up and Mr Brown retires below or paces the deck above a row of upturned disappointed face. Ere long his is listening over the side to the stories and appeals of individuals, some of whom may be invited on board to talk over their cases. A supplementary distribution often ensues, the donations being dropped down into the boats – sometimes, it is said between the boats, the yacht owner pretending to share in the consternation of the suitor as the gift, or part of it, sinks from sight, but invariably making the loss good. At last a halt is called and the afternoon visitors depart, as a rule; but unsatisfied applicants sometimes cling on for hours, joined later by the contingent of 'locals' who attend only at night, and who often have prolonged interviews with the owner on boat till midnight or even the small hours, That some of the applicants seem to think they have acquired prescriptive right to Mr Brown's bounty is evidenced by their noisy explosions of impatience now and then. They even venture upon remonstrance's, not very mildly expressed – a procedure which usually appears to amuse the unconventional philanthropist.

Persons ignorant of what constitutes a legal claim have actually sued to compel payment of money gifts alleged to have been promised, one actually obtaining judgement through the proceedings being undefended, so that the millionaire, to the great surprise of the district, was afforded the unwonted experience of a couple of years ago of having the county court bailiffs in temporary possession of his yacht!

It will be apparent that the methods of Mr Brown's benefactions are not exactly in consonance with those customarily adopted by practical philanthropists. Nor is he a great supporter of charities through the usual channels, although several local clergymen receive sums from him annually for the poor of their parishes. For this some, at least go out like the other applicants, as it is understood that no notice is taken of written appeals, and so a vicar of the Church of England may be seen among those who patiently wait in boats to see the millionaire.

The protracted waits now frequently imposed on many who formerly receive their gifts almost immediately they arrive is locally regarded as a sign that Mr Brown is tiring of the everlasting attentions of those who diurnally watch his movements and shiver by the hour at his gangway. Lately many of the regular participants in his bounty have attended for five consecutive days before receiving anything for the week. In that period some of the women occupied fifty to sixty hours, and walked over fifty miles in quest of one week's dole.

With whatever feeling the suppliants originally embarked in this enterprise, there is very little diffidence exhibited nowadays, although the promptitude with which attempts to photograph the assembled boats is resented points its own tale, The spectacle presented by the stream of importunate suitors is not one comforting to Anglo-Saxon pride and the violent rush for his bounty can scarcely have enhance whatever opinion of the average Briton Mr Brown entertained at the time of his arrival. On the other hand, there is a firm conviction among the more discriminating in the district that the exercise of the millionaire's benevolence in the manner described has had a detrimental local effect, and that the status and prospects of many of his pensioners, and of their families, will not be ultimately benefited by what may at present appear to be their good fortune.

During recent years the owner of the *Valfreyia*, it is understood, has received more than one large bequest, under wills of relatives, and he is reputed to be a millionaire several times over. Formerly he went somewhat frequently to London and other places on business and social visits, also walking and driving in the neighbourhood. But now though occasionally going to London for a day, he rarely leaves his yacht, sometimes not even going ashore for months. The only other occupants of the steamer being the skipper and a dozen hands (local men, periodically of duty ashore), possibly the extraordinary pursuit maintained by the people provides not altogether unwelcome interruptions to solitude which might, perhaps, be appreciated by the student, writer, painter, or inventor – to none of which classes, however, does this eccentric recluse appear to belong.

Temple, F.R., 'A Floating Goldmine', *Wide World Magazine*, May 1907 (CUL, L900. *c.*1928).

Endnotes

1. E. Dickin, *The History of Brightlingsea* (Brightlingsea, 1939), 3, 10.

2. Essex Record Office Publications, No.30, *Essex and the Sea* (Chelmsford, 1970), Fig.22.

3. 1851 Census of Great Britain, Population Tables 2, Table [1], 'Population Abstract'.

4. J. Wilson, *The Imperial Gazetteer of England and Wales* (Edinburgh, 1874), 1, 270.

5. Today Brightlingsea is a busy maritime town with some remaining commercial sea trade, but essentially a centre for the yachting and pleasure-boat industry. A substantial re-development of the old shipyards on the waterfront is currently underway.

6. Extracts/transcripts, Brightlingsea (1700–1909), ERO/COL-T/P 1/15.

7. H. Benham, *Essex Gold* (Chelmsford, 1993), Ch.10–11.

8. R. Neild, *The English, The French and the Oyster* (1995), 12.

9. Benham, 99–105.

10. Benham, 89–93.

11. Miscellaneous (1420–1812), ERO-T/P 1/14

12. C. Dove, *The Liberty of Brightlingsea* (Brightlingsea, 1974).

13. F. Youngs, *Guide to the Local Administrative Units of England*, 1 (1979), 132.

14. E. Ravenstein, 'The Laws of Migration', *Journal of the Statistical Society of London*, 48, 2 (June 1885), 167–235.

15. J. Long, 'Rural-Urban Migration and Socioeconomic Mobility in Victorian Britain', *The Journal of Economic History*, 65, 1 (March 2005), 1–35.

16. Dickin, *Brightlingsea*.

17. In particular, Benham, *Essex Gold*, and J. Leather, *The Northseamen* (Lavenham, 1971).

18. H. Hall, *Lucullus: or Palatable Essays*, 1 (1878), 7.

19. D. Starkey (*et al.*), *England's Sea Fisheries* (2000), 43, 90.

20. C. Dickens, *The Posthumous Papers of the Pickwick Club*, 1, (London Edition, 1837), 229.

21. J. Drummond, *The Englishman's Food* (1939), 366.

22. Hall, 18.

23. Drummond, 228.

24. P. Quennell, (ed.), *Mayhew's London* (1949), 100–101.

25. Hall, 12.

26. H. Mayhew, *London Labour and the London Poor*, in C. Yonge, *Oysters* (1960), 155 (Yonge's calculation of £125,000 appears to be erroneous).

27. Admission register (1883–97), ERO-D/Q 40/2; oyster company files (1882), NA-BT 31/3022/17094.

28. No records of wages have been uncovered, but J. Leather notes in *The Northseamen* (Lavenham, 1971), 36, that a smack could earn £12 per week as a fish carrier for larger trawlers.

29. For a full description of the evolution of the oyster dredge see H. Benham, *Essex Gold* (Chelmsford, 1993), 14–17.

30. Leather, 70–1.

31. Constructional details come from the study of surviving second-class smacks and from Peter Allen formerly of Brightlingsea and Alan Williams of Saint Osyth Boatyard, both of whom have considerable experience in Colne smack restoration.

32. Leather, 28–36.

33. R. Marks (*et al.*), *Pioneer, Last of the Skillingers* (Wivenhoe, 2002), 22–3.

34. White's *Directory of Essex* (1848), 450–1. It is unclear if the 160 smacks mentioned were all from Brightlingsea. Most of the town's mariners worked the common grounds near the harbour or went offshore; see Benham, 84–5.

35. Mr Pennel's report in *House of Commons Accounts and Papers*, LXIV (Shipping) session (1867).

36. Leather, 29.

37. Details taken from *BPM*; Mercantile Navy List (1883) and Fishing Register (1893) as quoted in Marks (*et al.*), 84.

38. *Pioneer* (CK88), the last surviving first-class east coast smack, was built by Peter Harris of East Donyland in 1864, and lengthened and adapted by Aldous of Brightlingsea in 1889. She is currently preserved and sailed by the Pioneer Sailing Trust; see www.pioneersailingtrust.org.uk. See also Leather, 304–5.

39. Original copy of this map loaned by Mr Guy Smith.

40. Building plans (1912–15), ERO/COL-D/UBR PB23/3; plan, Brightlingsea pier (1867), ERO-D/DU 210/36.

41. Certificates of ordination (1861–72), ERO-D/P 312/2/1.

42. The *Brightlingsea Parish Magazine (BPM)* carried a regular feature each month, 'Gossip from the Hard': the 'hard' being the waterfront area where smacks laid-up.

43. 'Gossip from the Hard', *BPM* (January, 1884).

44. Benham, 92.

45. *BPM* (July, 1884); Benham, 86; E. Dickin, *History of Brightlingsea* (Brightlingsea, 1939), 97–8.

46. H. Benham, *Last Stronghold of Sail* (1948), 33; Harbour reports (1865–1917), NA-MT 10/186. Stow-boating was the practice of rigging a net and 'stalling' the vessel into the tide, see Glossary.

47. *BPM* (May-September, 1900); see also Leather, 301–3.

48. *BPM* (September, 1884).

49. PP(O/L): 1808 (296) Bill for more effectual protection of oyster fisheries; 1823 (461) Bill for protecting and regulating public oyster fisheries in England and Wales; 1876 (65) Bill for better protection of oyster fisheries.

50. J. Hore and E. Jex, *The Deterioration of the Oyster and Trawl Fisheries of England: Its Cause and Remedy* (1880), 50.

51. *BPM* (January, 1895), see also R. Neild, *The English, The French and The Oyster* (1995), 105.

52. *BPM* (January–May, 1900).

53. *BPM* (February, 1905).

54. *BPM* (April, 1905).

55. *BPM* (October–December, 1910).

56. Benham, *Essex Gold*, 93.

57. *BPM* (September, 1914).

58. E. Dickin, *History of Brightlingsea* (Brightlingsea, 1939), 53.

59. Court book (1826–39), ERO-D/DU 210/10.

60. Court book (1864–80), ERO-D/DU 210/14.

61. D/DU 210/14.

62. From cross reference with census records.

63. D/DU 210/14.

64. A. Cross, 'Old English Local Courts and the Movement for their Reform', *Michigan Law Review* (1932), 30, 3, 371.

65. Select vestry orders, relief of poor (1827–8), ERO-D/P 312/8/1.

66. Names have been cross-referenced with census data.

67. Indentures (1653–1833), ERO-D/P 312/14/1.

68. Consecration, chapel-of-ease (1836), ERO/COL-D/P 8/1/16.

69. Minute books (1842–75), ERO-D/P 312/8/4.

70. Sequestration order (1878), ERO-D/P 312/7/1.

71. D/P 312/8/4.

72. Certificates of ordination (1861–72), ERO-D/P 312/2/1.

73. Dickin, 97–8.

74. www.cinqueportliberty.co.uk, accessed May, 2007.

75. F. Young, *Local Administrative Units of England* (1979), 132.

76. Deed of gift, chain of office (1892), ERO-D/DCm/Z1A.

77. C. Dove, *The Liberty of Brightlingsea* (Brightlingsea, 1974); Dickin, 117. Before the Norman Conquest, King Edward had contracted the five most strategically placed Channel ports to provide ships and men for royal service. Under the Anglo-Norman kings this became the principal means of keeping the two halves of their realm together, but after the loss of Normandy in 1205, the ships (the forerunners of the Royal Navy) ultimately became England's first line of defence. Brightlingsea became a Limb of the Head Port of Sandwich, contributing to that town's ship-service quota and providing a useful half-way stop en route to the annual Herring Fair at Yarmouth. For the Lord Warden, it extended his powers north of Sussex and Kent over the full width of the Thames estuary and provided good oysters: he had his own layings in Brightlingsea Creek until around the 1670s. Today the Cinque Ports have no more than a ceremonial role, although a base for the Lord Warden of the Ports is still maintained at Walmer Castle and new Lords Warden are always installed at Dover. All members of the Confederation, together with their Limbs, are situated in Kent and Sussex, except for Brightlingsea.

78. Young, 132.

79. Dove, *Liberty*, and www.brightlingsea-town.co.uk/history, accessed February–March 2007.

80. Dickin, 53.

81. *Crockford's Clerical Directory* (1888).

82. Extracts/transcripts, Brightlingsea (1700–1909), ERO/COL-T/P 1/15.

83. Prisoners accounts, county gaol (1828–34), ERO/COL-D/P 312/18/7.

84. CUL: House of Commons Accounts and Papers, V.16, (1854), Vol. LIV, 12. Crime statistics for Essex in 1853 show that 80 per cent of persons committed for trial were accused of non-violent offences against property.

85. M. Lockwood, 'Marine Policing in Essex', *Essex Police History Notebook*, Pt.1, No.31.

86. D. Hey (ed.), *The Oxford Companion to Local and Family History*, (1996), 328–9.

87. A. Everitt, 'The Pattern of Rural Dissent: the Nineteenth Century', *University of Leicester Dept. of English Local History Occasional Papers*, 2nd Series, 4 (1972), 17, 33.

88. Chapel book (1822), ERO/COL-D/NM 1/2.

89. J. Baker, *Introduction of Wesleyan Methodism into Brightlingsea* (1844), ERO Transcript 295; Accounts (1822–57), ERO/COL-D/NM 1/1.

90. D/NM 1/2; Plans for Wesleyan school (1861), ERO/COL-E/P 15/1.

91. A. Wakeling, *Brightlingsea Society of the New Church: A History 1813–1968* (1968), 8.

92. Quarter sessions, registered place of worship (1840–1), ERO/COL-Q/SBb 542/20.

93. Wakeling, 11–25.

94. D. Steel, *Sources for Nonconformist Genealogy and Family History* (1973), 787.

95. Q/SBb 542/20.

96. Church book (1846–94), ERO/COL-D/NC 30/1, cross referenced with census data.

97. *Essex Standard* (30 March, 1864); Dickin, 100.

98. Church circular (1889), ERO/COL-D/NC 30/19.

99. Congregational church register (1891–1904), ERO/COL-D/NC 39/16.

100. Dickin, 101.

101. Plan for Salvation Army hall (1908), ERO-D/UBR PB21/15.

102. PP(O/L), 1819 (224) Select Committee on Education of the Poor (1818), Digest of Parochial Returns, 253.

103. Rural Dean's parochial returns (1844), ERO-D/P 138/28/2.

104. PP(O/L), 1862 Committee of Council on Education: report, appendix (1861–1862), 510.

105. PP(O/L), 1870, Department of Science and Art: 17th Report, Appendix, 168; 1875, Committee of Council on Education: Report, Appendix, (1874–5) bye-laws of school boards.

106. PP(O/L), 1882, Committee of Council on Education: Report, Appendix, 404.

107. *London and Provincial Medical Directory* (1854–1892).

108. *Essex Standard* (1841–2), ERO-T/A 572/7.

109. Dickin, 264.

110. *Ancient Order of Foresters Directory*, London (1967), see also Dickin, 235.

111. *BPM* (September 1884).

112. www.brightlingsea-town.co.uk, accessed May, 2007.

113. E. Higgs, *A Clearer Sense of the Census* (1996), 143–68.

114. Population data for areas other than Brightlingsea represented in charts in this chapter are derived from census abstracts via www.visionofbritain.org.uk, B. Mitchell, *Abstract of British Historical Statistics* (Cambridge, 1962), and cross-checked with census abstracts as detailed in the bibliography.

115. Higgs, 147.

116. M. Anderson in E. Wrigley (ed.), *Nineteenth Century Society* (Cambridge, 1972), 55.

117. D. Baines, *Nineteenth Century Society*, 311.

118. M. Anderson, *Nineteenth Century Society*, 55.

119. W. Armstrong in H. Dyos (ed.), *The Study of Urban History* (1968), 67, 68.

120. F. Youngs, *Guide to the Local Administrative Units of England*, 1 (1979), 132.

121. C. Pooley and J. Turnbull, *Migration and Mobility in Britain Since the 18th Century* (1998), 93–95.

122. C. Elrington (ed.), *Victoria County History of Essex*, 9 (1994), 177–185.

123. Young, 157.

124. Data from 1841, 1851, 1861, 1871 and 1881 Census.

125. Oysters under one year old (see Glossary).

126. H. Benham, *Essex Gold* (Chelmsford, 1993), 97.

127. J. Wilson, *Imperial Gazetteer of England and Wales* (Edinburgh, 1874), 1, 516–7.

128. Wilson, 2, 1029 (20 fishing boats, 76 small and 73 large sailing vessels, and 3 small steam vessels belong to Wells in 1864).

129. Wilson, 2, 1062.

130. Wilson, 1, 281.

131. T. Lummis, *Occupation and Society* (Cambridge, 1985), 18–21.

132. Although these apprentices have no specific trade shown in the census, many lived with mariners.

133. Tide waiters waited for vessels coming in, to collect duty on goods (see Glossary).

134. E. Higgs, *Making Sense of the Census: The Manuscript Returns for England and Wales, 1801–1901* (1989), Ch.11; Armstrong, *Nineteenth Century Society*, Ch.6.

135. D. Gordon, *A Regional History of the Railways of Great Britain*, 5 (Newton Abbot, 1990), 68.

136. E. Wrigley & R. Schofield, *The Population History of England, 1541–1871* (Cambridge, 1989), 220.

137. London and Essex, Kent, Middlesex and Surrey.

138. E. Hunt, *Regional Wage Variations in Britain: 1850–1914* (Oxford, 1973), 226.

139. R. Woods, *The Demography of Victorian England and Wales* (Cambridge, 2000), 383.

140. Pooley and Turnbull, 11–19.

141. E. Ravenstein, 'The Laws of Migration', *Journal of the Statistical Society of London*, 48, 2 (June 1885), 167–235.

142. J. Corbett, 'Ernest George Ravenstein: The Laws of Migration, 1885', www.csiss.org/classics, accessed June 2006.

143. J. Long, 'Rural-Urban Migration', *Economic History*, 65, 1 (March 2005), 1–35.

144. G. Nair and D. Poyner, 'The Flight from the Land? Migration in South-East Shropshire in the Late Nineteenth Century,' *Rural History*, 17, 2 (2006), 167–186.

145. Benham, 97–8.

146. H. Dyos, *Victorian Suberb: A Study of the Growth of Camberwell* (Leicester, 1961), 19.

147. W. Ashworth, *The Genesis of Modern British Town Planning* (1954), 81.

148. Dyos, 51.

149. Taken from CUL/Atlas.2.97.6., plates xiv and xv.

150. Brightlingsea Conservation Area – final report, *Tendring District Council Conservation Area Review* (2006), 3.

151. www.brightlingsea-town.co.uk/history, accessed March, 2007.

152. T.D.C., Report, 3.

153. Tithe map (1841), ERO/COL-D/CT 51B.

154. T.D.C. Report, p.3.

155. Sale catalogues (1795–1886), ERO/COL-D/DEL E26.

156. Building plots for sale, indentures (1828–60), ERO/COL-D/DCfT51.

157. D/DCfT51.

158. John Dorrien Magens was also responsible for the connection of East Grinstead to the railway at Three Bridges in 1855 (www.hammerwood.mistral.co.uk, accessed February 2007).

159. D. Gordon, *A Regional History of the Railways of Great Britain,* 5 (Newton Abbot, 1990), 68.

160. D/DCfT51.

161. 6in O.S. map (1875).

162. Building plans (1907–15), ERO/COL-D/UBR PB23/3 and ERO/COL-D/UBR PB21/4.

163. O.S. maps for Brightlingsea (1897, 1923).

164. B. Mitchell, *Abstract of British Historical Statistics* (Cambridge, 1962), 236–7.

165. 1881 Census.

166. T. Sharp, *Town and Townscape* (1968), 9.

167. Prior to the advent of large 'out of town' supermarkets. Brightlingsea now has one nearby.

168. O.S. map (1897); Sale of shipyards, buildings and land (1886), ERO-ACC. C. 32.

Other titles available from The History Press:

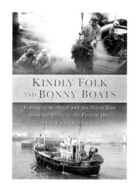

Kindly Folk and Bonny Boats: Fishing in Scotland and the North-East from the 1950s to the Present Day
GLORIA WILSON

This book provides a pictorial appreciation of the boats and fishing communities of Scotland and North-East England from the 1950s to the present, making use of Gloria Wilson's unique collection of photographs. It includes information on boat design and construction as well as some rarely seen naval architects' line plans.

978 0 7524 4907 4

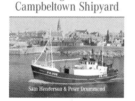

Fishing Boats of Cambeltown Shipyard
SAM HENDERSON AND PETER DRUMMOND

Competition from foreign shipyards moved into a new dimension from the mid-nineties onwards. In addition, decommissioning, restricted days at sea and shrinking quotas have left the Scottish fleet a shadow of its former self. However, by the beginning of the twenty-first century, things were beginning to look up for the remaining vessels, including surviving boats built in Cambeltown Shipyard.

978 0 7524 4765 0

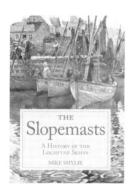

The Slopemasts: A History of the Lochfyne Skiffs
MIKE SMYLIE

The West Coast of Scotland has its own peculiarities that have led to altogether different boats developing through their usage. Amongst these craft is the Lochfyne skiff that emerged from several generations of innovation and which resulted in one of the prettiest workboats to have graced the British shores.

978 0 7524 4774 2

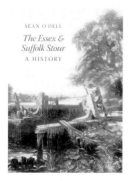

The Essex and Suffolk Stour: A History
SEAN O'DELL

This book will detail the beginnings, life and trading decline of the Stour Navigation from the seventeenth century to the present day, looking at the circumstances surrounding the construction of the first lock gates and general engineering works that converted the river into an inland navigation and the changing fortunes in the eighteenth and nineteenth centuries.

978 0 7524 3911 2